INTERMEDIATE
ESL VIDEO LIBRARY

INNOVATIONWATCH

Isis C. Clemente

Susan Stempleski
Series Editor

PRENTICE HALL REGENTS
Englewood Cliffs, New Jersey 07632

Acquisitions Editor: **Nancy Leonhardt**
Manager of Development Services: **Louisa Hellegers**
Development Editor: **Barbara Barysh**
Assistant Editor: **Sheryl Olinsky**
Editorial/Production Supervision: **Dominick Mosco**
Page Composition: **Lido Graphics**
Production Coordinator: **Ray Keating**
Cover Supervisor: **Merle Krumper**
Cover Design: **Marianne Frasco**
Cover Photograph: **Comstock**
Interior Design: **Function Thru Form**
Electronic Artist: **Todd Ware**

©1996 by PRENTICE HALL REGENTS
Prentice–Hall, Inc.
A Simon & Schuster Company
Englewood Cliffs, New Jersey 07632

Capital Cities/ABC Multimedia Group

Printed in the United States of America
10   9   8   7   6   5   4   3   2   1

ISBN 0-13-094038-0

Prentice-Hall International (UK) Limited, *London*
Prentice-Hall of Australia Pty. Limited, *Sydney*
Prentice-Hall Canada Inc., *Toronto*
Prentice-Hall Hispanoamericana, S.A., *Mexico*
Prentice-Hall of India Private Limited, *New Delhi*
Prentice-Hall of Japan, Inc., *Tokyo*
Simon & Schuster Asia Pte, Ltd., *Singapore*
Editora Prentice-Hall do Brasil, Ltda., *Rio de Janeiro*

Printed on Recyclable Paper

TABLE OF CONTENTS

ACKNOWLEDGMENTS

I would like to express my gratitude to a group of very special people whose encouragement assured the completion of *InnovationWatch*:

First and foremost to Susan Stempleski, co-author and editor of this series, without whose infinite patience, guidance and praise, *InnovationWatch* would have never been completed; to Barry Tomalin, former Editor and Head of Marketing for BBC English and a co-author of this series, for having faith in me; to Nancy Leonhardt, Acquisitions Editor at Prentice Hall Regents for giving me the opportunity to work on the project; to Barbara Barysh, ESL/EFL Development Editor, for her valuable suggestions and painstaking editorial work; and to the entire editorial and production staff at Prentice Hall Regents for successfully completing the overwhelming task of turning a manuscript into a book.

At Miami-Dade Community College, North Campus, the support during hard times was inspiring and motivational. Thanks to Leslie Biaggi, Karen Bryant, Carol Call, Kelly Cardona, Gina Cortes-Suarez, Rick Parendes, Juan Perez, Greg Sharp, and Maria Vargas.

On a more personal note, thanks to both of my children, Erik and Ector, whose endless questions never allow me to forget the child in me. Last, but not least, a very, very special thanks to Lloyd Madansky whom I consider my greatest critic and my biggest fan.

I would like to dedicate this book to an immigrant couple who worked very hard to make it in a new country and culture and were able to accomplish many of the dreams they had for their children:

To my parents:

Antonio and Valentina Clemente

INTRODUCTION TO THE SERIES

The *ABC News Intermediate ESL Video Library* is an interactive, integrated skills series designed for intermediate level adult learners of English as a second or foreign language. The series consists of five videocassettes: *BusinessWatch, CultureWatch, EarthWatch, HealthWatch,* and *InnovationWatch,* each accompanied by a student text and Instructor's Manual.

THE VIDEOS

Each videocassette consists of twelve actual broadcast segments from ABC News programs such as *World News Tonight, 20/20, Prime Time Live, The Health Show,* and *Business World.* These authentic television news reports focus on high-interest topics and expose students to natural English, spoken by a wide variety of people from diverse backgrounds and age groups. The videos are time coded and the codes are given in the textbooks so that sequences may be easily identified. The videos are also closed captioned. Teachers who have access to a closed caption decoder may wish to have their students view the captions as they carry out some of the activities. Some suggestions for activities based on closed captions are described at the end of this introduction.

THE BOOKS

Each book offers a broad range of task-based activities centering around the selected video segments. These activities provide practice in all four language skills: listening, speaking, reading, and writing. The reading material parallels or extends the news story and is drawn from several sources. The books also contain complete transcripts of the video segments. The Instructor's Manual, for each book contains an Answer Key to all the activities.

The general aims of the books are:
* To enhance comprehension of each video segment.
* To highlight and exploit specific language on the video.
* To stimulate discussion about the topics presented on the video.
* To offer authentic reading material related to the content of the video.
* To give students practice in writing clear and simple English.

SOME GENERAL SUGGESTIONS FOR TEACHING WITH VIDEO

Only recently has video moved from being something that is switched on and left to present language without the teacher's intervention to becoming a flexible resource for classroom activities. While there is no one "right way" to use video in language teaching, teachers using the materials in the *ABC News Intermediate ESL Video Library* will probably find the following general guidelines helpful:

Familiarize Yourself with the Material. Before presenting a lesson in class, view the entire sequence yourself, preferably several times and with the video transcript in hand. If time allows, try doing the activities yourself, in order to anticipate difficulties or questions your students may have.

Allow for Repeated Viewing. In order to carry out the viewing activities in the lessons efficiently, students will need to see and hear the video sequence and selected portions several times. Each viewing activity in the lessons is accompanied by a time code. Refer to these time codes and then play or replay the indicated section of the video in conjunction with the particular viewing task at hand.

Present Activities to the Students Before Viewing. Students will focus their attention more effectively on the viewing activities if you ensure that they understand the directions for each task before playing or replaying the video sequence.

Get to Know Your Equipment. Practice with the video equipment you will be using in class. The time codes on the video will help you locate the sequence to be shown and any other points you may wish to highlight.

HOW TO USE THE VIDEOS AND THE BOOKS

You can use the news stories in the *ABC News Intermediate ESL Video Library* in the order in which they are presented on the videos and in the books, or you may choose particular video segments according to the interests of your students. The video segments are not graded in terms of grammatical difficulty, and there is no artificial variation in linguistic complexity from lesson to lesson within the books. You can have students work through each exercise in a lesson, or you may choose specific activities to suit your students' needs and particular class schedule.

Each book in the *ABC News Intermediate ESL Video Library* consists of twelve 10-page lessons, each corresponding to a single news segment on the video. Every lesson is structured in the same way and has three main sections: BEFORE YOU WATCH, WHILE YOU WATCH, and AFTER YOU WATCH.

BEFORE YOU WATCH

This section contains previewing activities that prepare students to watch the video by tapping their background knowledge and stimulating interest in the topic. There are three types of activities in this section:

- **Talking Points:** These are questions designed to stimulate general discussion and elicit relevant vocabulary and background knowledge about the topic of the video segment. They also motivate students to watch the video and provide opportunities for them to exchange ideas.

- **Predicting:** These activities encourage students to think about the topic and to predict the kinds of information they think will be included on the video.

- **Key Words:** This is a vocabulary activity to introduce or review words directly related to the topic of the video.

WHILE YOU WATCH

This section contains a variety of viewing activities for students to complete while actually watching the video. The activities promote active viewing and listening and facilitate comprehension by focusing on essential features of the news story. Time codes corresponding to the appropriate section of the video are printed next to each activity. These time codes facilitate access to the relevant part of the video to be played or replayed in conjunction with the particular viewing task.

- **Getting the Main Idea:** This is a global viewing activity in which students watch the entire video segment and answer questions about key ideas (the who, what, where, why, when, and how) of the segment.

The remaining activities in the WHILE YOU WATCH section take students through the video bit by bit to focus on more specific information. In order to carry out these activities efficiently students are asked to watch and hear particularly relevant sections of the video again to gain a detailed understanding of the news story. The activities used and the order in which they are presented vary, but they include a combination of several of the following types:

- **Checking Your Predictions:** Students watch the video and check to see if their predictions about the content of the video were correct.

- **What's Missing?:** This is a listening cloze. Students listen to a portion of the video segment and fill in the missing words in a section of the video transcript.

- **True or False?:** Students watch a section of the video and indicate whether the statements in the exercise are true or false. Students are additionally challenged by being asked to change the false sentences to make them true.

- **Checking What You Hear:** Students watch the video and check particular details that are mentioned on the video.

- **Listening for Details:** Students watch the video and circle the correct answers to a series of multiple-choice type questions.

- **Checking What You See:** This activity encourages students to pay close attention to visual information presented on the video. Students watch the video and check the images they actually see.

- **Notetaking:** Students watch the video and take brief notes on the answers to a series of *Wh-* questions focusing on specific details.

- **Information Match:** This is a matching exercise. Students watch the video and match the names of people, places, or things with related information.

- **Putting Events in Order:** This is a sequencing activity. Students watch the video segment and number a series of events in the order in which they are presented on the video.

- **Making Inferences:** Students watch the video and are asked to reach logical conclusions, based on the facts presented in the news story.

- **Identifying What You See:** Students watch selected portions of the video and identify what they see by writing a brief description.

- **Making True Sentences:** This is a matching exercise in which students write true sentences about the video by combining sentence stems and endings.

- **Who's Who?:** Students watch the video and check the sentences that apply to different people who are interviewed or shown on the video.

AFTER YOU WATCH

This section contains postviewing activities related to the topic of the video. The activities stimulate language use and encourage students to integrate information from the video. This section always contains the following six activity types:

- **Language Point:** In this activity, a selected language function or grammar point from the video is highlighted for further practice.

- **Vocabulary Check:** This activity reviews idioms, colloquial expressions, and other words or phrases used on the video.

- **Discussion:** These are questions that encourage students to relate the information on the video to their own lives and/or the situation in their own countries.

- **Role Play:** This activity provides students with an opportunity to use the information and language they have acquired while working on the video in a freer practice stage.

- **Reading:** This activity offers authentic reading material — such as magazine articles — related to the content of the video.

- **Writing:** In these activities, students are encouraged to integrate and use information from the video to prepare letters, short news articles, and other original documents.

SOME SUGGESTIONS FOR USING CLOSED CAPTIONS

As mentioned above, an additional feature of the videos in the *ABC News Intermediate ESL Video Library* is that all the video segments, with the exception of those from *The Health Show* and *Business World*, are closed captioned. If you have access to a closed captioned decoder (or a television with the new decoder chip) you may wish to open up the captions and have your students do the following variations of some of the activity types in the books:

- **Key Words:** Have students read the captions and call out "Stop!" when they see one of the key words in the captions. Use the pause button to stop the video at that point, and have students suggest the meaning of the word as it is used in the segment.

- **Checking Your Predictions:** Instead of having students *listen for* the information which they have predicted, they can be asked to *read* the captions to locate the relevant information.

- **What's Missing?:** As a variation of this activity, students can be asked to fill in the blanks (using any words that make sense) *before* watching the video. Students then watch the video, and read the captions to check their answers.

- **Vocabulary Check:** As in the variation of the Key Words activity just described, students can be asked to read the captions and indicate when they *see* the idioms, colloquial expressions, and other words or phrases highlighted in the Vocabulary Check.

These are just a few ways in which the closed captions on the segments can be used to enhance language learning. Allow your imagination to come up with ideas. In situations in which a closed caption decoder is not available, students can carry out similar activities by using the video transcripts in the back of the book.

SUGGESTIONS FOR STUDENTS WORKING WITHOUT A TEACHER

Language learners differ from one another in many ways. If you are learning English without a teacher, you should use the materials in the *ABC News Intermediate ESL Video Library* in the way that is most suitable to *you* and *your* situation. However, your work will probably be more pleasant and productive if you do the following:

- Follow the plan of each lesson.

- Read the directions to each exercise carefully.

- Use the Answer Key only when you have to — and that should be after you have completed the exercise. The Answer Key is in the Instructor's Manual.

- If an exercise does not have clear-cut answers in the Answer Key, try to do the exercise with another person: a native speaker or someone at your own level of English.

- Use the transcripts of the video segments to study the language used on the video in detail. The transcripts are printed in the back of the book.

- Set realistic goals for yourself as you work on the lessons. As with learning many other things, the key to successful language learning is to do a small amount of work regularly and frequently. If doing one lesson a week is too much, try doing one over two weeks.

- If you have a closed caption decoder, you may find it helpful to follow along with the speakers and read the words on the screen.

Finally, enjoy improving your English with the *ABC News Intermediate ESL Video Library!*

Segment 1
What Are the Differences Between Men and Women?

From: *Nightline*, 7/31/91
Begin: 00:32
Length: 6:54

BEFORE YOU WATCH

TALKING POINTS

Work in groups. Discuss your answers to the following questions.

1. Are there differences between men and women? Make a short list of what you consider to be the vital differences.

2. Do you believe that men and women have the same intellectual capacity? Give examples to support your answer.

3. Should men and women hold the same kinds of jobs at home? in the business world? in education? in government? in the military?

4. Do you believe one of the sexes is superior to the other? If so, which one? Explain your answer.

PREDICTING

Work in groups. Based on the title of the video and the picture above, predict the kinds of information you think will be included on the video.

1. _____

2. _____

3. _____

KEY WORDS

The *italicized* words in the sentences below will help you understand the video. Study the sentences. Then match the words with the meanings.

1. There are two major *sexes* in plants and animals.
2. "Homo sapiens" is the scientific name for the human *species*.
3. The process of *evolution* is still going on.
4. He has a very nervous *temperament*.
5. Mary is *indistinguishable* from her twin sister.
6. The doctor said the eight-month-old *fetus* appeared to be normal.
7. Medical tests showed that he had a very high level of *testosterone*.
8. She's *altering* her clothes because she has lost weight.
9. Males have a tendency to be more *aggressive* than females.
10. Kindness and generosity are her best *traits*.

1. _____ *sexes*	a. not easily recognized as different
2. _____ *species*	b. an unborn animal or human being
3. _____ *evolution*	c. particular qualities of a person
4. _____ *temperament*	d. a distinct group of plants or animals
5. _____ *indistinguishable*	e. threatening or ready to attack
6. _____ *fetus*	f. sets of male and female creatures
7. _____ *testosterone*	g. making changes in
8. _____ *altering*	h. a male hormone
9. _____ *aggressive*	i. a person's basic personality or nature
10. _____ *traits*	j. development of new forms of life from earlier, simpler forms

WHILE YOU WATCH

00:39–
07:33

GETTING THE MAIN IDEA

Watch the news report and listen for the answers to the following questions. Take brief notes on the answers. Then compare your answers with those of another student.

Who is doing **what, where?**

Why and **how** are they doing this?

Who?	
What?	
Where?	
Why?	
How?	

CHECKING YOUR PREDICTIONS

Look at your answers to the predicting exercises on page 1. Watch the video again and check (✓) the kinds of information that are actually included on the video.

00:39–
07:33

WHAT'S MISSING?

Listen to Forrest Sawyer's introduction to the news report. Fill in the missing words.

00:39–
00:55

Forrest Sawyer: It took the Gulf (1) _____ to make us notice just how much has changed between the (2) _____. Day after (3)_____ of the crisis, we saw men and women working side by side in the (4) _____. We saw fathers and mothers separated from their (5) _____, putting (6) _____ in harm's way.

TRUE OR FALSE?

Watch the next part of the news report again. Are the following statements *true* or *false?* Write **T** (true) or **F** (false). Make the false sentences true by changing one or two words.

00:39–
01:33

1. _____ Americans became aware of the changes between men and women after the Gulf War.

2. _____ During the crisis, men and women worked separately in the military.

3. _____ Only fathers were separated from their families.

4. _____ The U. S. Senate voted to allow female military pilots to fly combat missions.

5. _____ In the old days, it was fashionable to say that women were the gentler sex.

6. _____ Nowadays, we are sure that men are the superior sex.

CHECKING WHAT YOU HEAR

01:34–
02:39

Watch the video. What statements do you hear? Check (✓) the appropriate boxes.

1. ❑ Today's debate was carried out in a political arena.
2. ❑ Women do not have what it takes to be a soldier.
3. ❑ Males are not the superior sex.
4. ❑ What scientists are discovering might make politicians think differently about the differences between the sexes.

WHO'S WHO?

01:49–
02:19

Look at the chart below. Then watch the video and check (✓) the appropriate boxes.

Who . . . ?	William Roth	John McCain
1. is the Republican Senator from Delaware		
2. is the Republican Senator from Arizona		
3. believes that women have proven themselves		
4. thinks that this is not a women's rights issue		
5. says that women pilots have high performance, experience, and aptitude		

CHECKING WHAT YOU SEE

02:39–
03:36

Watch the next part of the video with the *sound off* and put a check (✓) next to the animals that you see.

1. ❑ monkeys   4. ❑ cows   7. ❑ deer   10. ❑ birds
2. ❑ dogs   5. ❑ elephants   8. ❑ cats   11. ❑ horses
3. ❑ snails   6. ❑ bees   9. ❑ seals   12. ❑ wolves

Watch the video again with the sound *turned up*. Which animals did you identify correctly?

NOTETAKING

Watch the next part of the video and take brief notes on the answers to the following questions. Then compare your notes with those of another student.

03:36–
05:16

1. Name the similarities between male and female fetuses in the early stages of development.

   _____

2. When do differences occur between male and female fetuses? Why?

   _____

3. What hormone is responsible for development of male body structure? What does this hormone do?

   _____

4. Name the consistent physical sex differences that Michael Guillen points out between males and females.

   _____

5. According to Ms. Hines, a neurobiologist, how are males and females more similar than different?

   _____

MAKING TRUE SENTENCES

Watch the video. Then match the sentence parts to make five true sentences. Write the sentences below. The first one has been done for you.

05:17–
05:58

1. Females	. . . 105 males are born.
2. For every 100 females born,	. . . there are 72 males.
3. By middle age, for every 98 males,	. . . live longer than males.
4. By old age, for every 100 females,	. . . committed by males than by females.
5. More murders are	. . . there are 100 females.

1. _Females live longer that males._ _____

2. _____

3. _____

4. _____

5. _____

LISTENING FOR DETAILS

Watch the last part of the video. Circle the correct answer.

05:59–
07:33

1. What happens when you inject female rats with testosterone?
   a. Their behavior remains the same.
   b. They become more aggressive.
   c. They become unconscious.

2. Why is it difficult to draw conclusions from the study that involved injecting female rats with testosterone?
   a. Because no male rats were involved in the study.
   b. Because female rats are very different from male rats.
   c. Because rats are very different genetically from humans.

3. According to a study done of Vietnam veterans, what were soldiers with high testosterone levels most likely to do?
   a. Go AWOL or end up in the brig.
   b. Be courageous.
   c. Be more intelligent.

4. According to Mr. Deacon, why aren't sex differences a good measure of the differences between people?
   a. There are too many people to test.
   b. We don't have a good instrument to use.
   c. There is so much individual variety.

5. What is the average height of men and women?
   a. Women are 64 inches tall. Men are 69 inches tall.
   b. Men are 64 inches tall. Women are 69 inches tall.
   c. Men are 66 inches tall. Women are 5 inches tall.

6. What does Michael Guillen conclude?
   a. There are no significant differences between the sexes.
   b. One sex is still proven to be superior to the other.
   c. Politicians have to investigate this further.

AFTER YOU WATCH

LANGUAGE POINT: COMPARATIVE FORMS OF ADJECTIVES

On the video, Michael Guillen uses *comparative forms of adjectives* when he says "In adulthood, the average male is *bigger* than the average female" and "the average female has *better* senses." Complete the following sentences with the comparative form of the appropriate adjective in the box.

high	good	long
large	tall	small

1. The average man is five inches _____ than the average woman.
2. Women have _____ lungs than men.
3. Women have a _____ sense of smell than men.
4. Women live _____ lives than men.
5. On the average, men have _____ muscles than women.
6. The number of murders committed by men is _____ than the number of murders committed by women.

VOCABULARY CHECK: COLLOQUIAL EXPRESSIONS

The following excerpts are from the video. What do the *italicized* expressions mean? Circle the correct answer.

1. We saw fathers and mothers separated from their families, putting themselves *in harm's way.*
   a. in a safe place
   b. in a dangerous situation
   c. a long distance away
   d. a short distance away

2. Today it joined the House in voting to allow military pilots to fly combat missions, which *stirs up an old hornet's nest.*
   a. prevents a problem
   b. solves a problem
   c. eliminates traditional jobs
   d. recalls a difficult situation

3. [An] age-old debate . . . *heated up* again today in the Senate.
   a. became angry and excited
   b. became calm and relaxed
   c. came to an end
   d. continued

4. Females live longer, by seven years, even though males start out *ahead of the game.*
   a. in a worse position
   b. in a better position
   c. older
   d. younger

5. I think what science has to say is that men and women *run the full gamut* of behavior and brain structure.
   a. show no possibilities
   b. show a few possibilities
   c. show many possibilities
   d. show all the possibilities

6. Increasingly, . . . the old idea that one sex is absolutely superior to the other is proving to be just a lot of *hot air.*
   a. worthless talk
   b. truth
   c. strong emotional feeling
   d. trouble

CATEGORIZING WORDS

The following words are used on the video. In which category does each word belong? Check (✓) the appropriate column.

	Hormones	Vital Organs	Sex Organs
1. heart			
2. lungs			
3. ovary			
4. testosterone			
5. testes			

DISCUSSION

Work in groups. Discuss your answers to the following questions.

1. What did you learn from the video?
2. Were you shocked or surprised by anything on the video? If so, what?
3. Did the video strengthen or confirm any of your own opinions regarding the differences or similarities between men and women? Explain your answer.

ROLE PLAY

Work in pairs. One student will play the role of the interviewer. The other will pay the role of the neurobiologist. Read the situation and the role descriptions below and decide who will play each role. After a ten-minute discussion, begin the interview.

THE SITUATION: **A Radio Interview**

A radio program called "Science Matters" has invited a neurobiologist to talk about biological differences and similarities between men and women.

ROLE DESCRIPTION: **Interviewer**

You are an interviewer for "Science Matters." Based on the information in the video, prepare a list of questions to ask the neurobiologist about the biological differences and similarities between men and women.

You are the neurobiologist. Be prepared to answer questions about the biological differences and similarities between men and women.

READING

Read the transcript of a discussion that a group of American college students had with their psychology professor. The purpose of the discussion was to draw out the personal opinions of the students about the differences between men and women. Answer the questions that follow.

Professor Wright: Much has been noted in your text about issues of equality between men and women. Today I'd like us to examine the other side of that coin. Are there differences in behavioral attitudes between men and women?

Nelson: No, I don't think there are any significant differences. Today's women have been raised in a pretty equal social and cultural environment to men. So, essentially they share all the benefits and stress-related problems that men face.

Deanna: Well, I disagree. Women are far from having achieved equal status. We see this in home life, the workplace, and in social settings. Although women have made advances in the past thirty years, male supremacy still prevails.

Bettina: You mean male supremacy prevails in social relationships! Certainly women are now as free as men to enjoy intimate relationships. The sexual revolution in the 1960s cleared the path for that freedom, although AIDS is doing a pretty good job of limiting the number of relationships one can enjoy. But that's true for both men and women.

Deanna: Well, even in relationships there are differences between the way a woman feels about her man and vice versa. A woman is more dedicated to her man. She looks at a relationship in a more holistic manner. If she is seriously involved in a relationship, everything she does is somehow connected to that relationship. If she buys something new for herself, she consciously thinks about whether it will please her man. If she makes travel plans, she includes him in them or works around his feelings so that he won't be hurt. A man doesn't think that way. He does what he's got to do and doesn't allow his relationship to interfere.

Nelson: Well, I object to that statement because it implies that men can't be as emotional as women. Men can be very emotionally tied into a relationship. I remember my first girlfriend. All I could do was think of her. I couldn't concentrate on anything else.

Bettina: That's just "being in love," Nelson. Everyone has those feelings, but they only last a short time.

Professor Wright: I'm not sure that Nelson has the same meaning for the word "short" that most of us have, Bettina!

Nelson: What do you mean by that?

Professor Wright: I mean that this is the last week in the semester, Nelson, and this is the first time you've paid attention in class. Have you broken up with your first girlfriend?

1. What have the students been reading about in their text?

   _____

2. What is the theme of this class discussion?

   _____

3. What is Nelson's view on the issue?

   _____

4. Which student clearly opposes Nelson's views?

   _____

5. Why was Professor Wright surprised at Nelson's performance in class?

   _____

6. Which student seems to have the most conservative point of view? Which seems to have the most liberal? Explain.

   _____

WRITING

1. Due to cultural upbringing, men and women may behave differently in social settings. Write about a situation in which a man might act one way but a woman would probably act differently.

2. Write a persuasive paragraph giving at least three reasons why women should not feel inferior to men. Include main idea, supportive ideas, and examples.

Segment 2
You Can Work It Out

From: *20/20*, 11/29/91
Begin: 07:34
Length: 7:26

BEFORE YOU WATCH

TALKING POINTS

Work in groups. Discuss your answers to the following questions.

1. Why do some marriages go bad? Describe some signs that indicate that a marriage is not going well.
2. If you had marriage problems, what would you do? tell your parents? tell your friend(s)? tell your children? ignore the problems and hope they would go away? talk to your spouse? seek professional help from a marriage counselor? leave the marriage? do something else?
3. How do your feel about seeing a marriage counselor or psychologist to discuss your personal problems?

PREDICTING

Work in groups. The video is about how some married couples try to solve their marriage problems. Write down three questions you think will be answered on the video.

1. _____

2. _____

3. _____

KEY WORDS

The *italicized* words below will help you understand the video. Study the sentences. Then write your own definition of each word.

1. They expected their marriage to be happy, but they were soon *disillusioned*.
   disillusioned: _____

2. Couples whose marriages are not going well may go to a *family therapist* for help.
   family therapist: _____

3. *Conflicts* between good friends should be resolved, not ignored.
   conflicts: _____

4. Jack and Nancy first met each other on a *blind date* that was arranged by a friend of theirs.
   blind date: _____

5. Husbands and wives often follow the *gender roles* that they have learned from their own mothers and fathers.
   gender roles: _____

6. Traditionally, the father is the *breadwinner* of the family.
   *breadwinner:*_____

7. My mother often *deferred* her own needs to those of her children.
   deferred: _____

8. They agreed on a *timetable* for the changes they wanted to make.
   timetable: _____

9. They had a hard time bringing their discussion to *closure*.
   closure: _____

10. His foolish behavior may put his whole future in *jeopardy*.
    jeopardy: _____

WHILE YOU WATCH

GETTING THE MAIN IDEA

07:42–
25:08

Watch the news report and listen for the answers to the following questions. Take brief notes on the answers. Then compare your answers with those of another student.

Who is this doing **what? Where?**

Why and **how** are they doing this?

Who?	
What?	
Where?	
Why?	
How?	

CHECKING YOUR PREDICTIONS

Look at the questions you wrote in the PREDICTING exercise on page 11. Watch the video. Which of your questions are answered on the video? What answers are given?

07:42–
25:08

WHAT'S MISSING?

Listen to Hugh Downs' introduction to the news report. Fill in the missing words.

07:42–
08:00

Hugh Downs: If you're (1)_____ , how is your marriage (2)_____ , and if it's not going well, do you (3)_____ enough to save it? A (4)_____ ago, (5)_____ seemed more (6)_____ to head right for the (7)_____ court, but now many marriage (8)_____ report more couples are trying to work it out. But exactly how do you do that, and isn't every marriage different?

LISTENING FOR DETAILS

Watch the next part of the video. Circle the correct answers.

08:13–
10:55

1. According to Stone Phillips, what question is perhaps "the single most perplexing" (the hardest one to answer) about marriage?
   a. Why do disillusioned people avoid marriage?
   b. Why do married couples become disillusioned?
   c. How can married people bring out the best in one another?

2. Which statement is NOT true about Judy and Bob Muller?
   a. They've been married for six years.
   b. They were both married before.
   c. They have two children.

3. When did Judy fall in love with Bob?
   a. On the night that she first met him.
   b. On the day he asked her to marry him.
   c. On the first night of their marriage.

4. Which statement is TRUE about Bob and Judy's early relationship?
   a. It was only a romantic attraction.
   b. They saw qualities they liked in each other.
   c. Bob thought Judy was wishy-washy (weak).

5. As Judy got to know Bob better, what kind of a husband did she think he would be?
   a. Close-minded, selfish, and unfair.
   b. Open-minded and driven to do what he wanted.
   c. Open-minded, generous, and fair.

6. What qualities attracted Judy and Bob to each another?
   a. Her independence and his open-mindedness.
   b. His independence and her open-mindedness.
   c. His generosity and her fairness.

7. How did Bob's behavior toward Judy change after they were married?
   a. He became independent and lazy.
   b. He became more open-minded and less controlling.
   c. He became less open-minded and more controlling.

8. How did Judy's behavior toward Bob change after they were married?
   a. She became less independent and let Bob make decisions.
   b. She became more independent and wouldn't let Bob make decisions.
   c. She ignored him and let her father make all the decisions.

10:56–
15:19

TRUE OR FALSE?

Watch the video. Are the following statements *true* or *false*? Write **T** (true) or **F** (false). Make the false statements true by changing one or two words.

1. _____ When two people get married, they often fall into gender roles.

2. _____ The man seldom takes the role of emotional support for the family.

3. _____ Men are better off letting women do the emotional work for the marriage.

4. _____ It is good to lose a sense of self in a relationship.

5. _____ When things don't go well, couples begin blaming each other.

6. _____ A woman should always defer her needs to the needs of her husband.

7. _____ Judy began to feel frustrated because she couldn't say "No" to her husband.

8. _____ In a marriage, the person who makes the money is the only one who should decide how to spend the money.

NOTETAKING

On the video, Jo-Anne Krestan, Stone Phillips, and Judy Muller each give an example of how men and women get locked into gender roles that can damage a marriage. Listen again to the video and take brief notes about the example each person gives. Then compare your notes with those of another student.

10:56–12:40

Jo-Ann Krestan: _____

_____

_____

Stone Phillips: _____

_____

_____

Judy Muller: _____

_____

_____

PUTTING EVENTS IN ORDER

Read the sentences below. Then watch the next part of the video and put the events in the correct order. Number them 1 to 5. The first event has been numbered for you.

15:20–24:47

_____ Bob realized that his father, Jack, also carried through the patterns that he learned from his father.

_____ Bob realized that he is not living his life according to his own needs.

_1_ Judy gave Bob an ultimatum.

_____ Bob decided to go see his father.

_____ Bob made an emotional connection with his father.

_____ Judy gave Bob a deadline of February 14th.

WHO'S WHO?

Watch the next part of the video. Match the names with the comments on marriage.

15:38–
25:06

   a. Stone Phillips   b. Judy Mueller   c. Judy Krestan

1. _____ I got tired of giving up my life all the time for his dreams.

2. _____ When one person changes in marriage therapy, the other person will soon either leave the marriage or change, too.

3. _____ Marriage therapy is like a mobile. After removing a little piece it's in desequilibrium and everything swings wildly until it finds a new balance.

4. _____ Patterns sort of filter down into marriages.

5. _____ As each partner takes responsibility for all aspects of their lives, they begin to grow again as individuals and as a couple in marriage.

INTERPRETING BEHAVIOR

Watch the parts of the video that show Judy and Bob crying. Then answer the following questions. Compare your answers with those of another student.

16:20–
18:40

1. Why is Judy crying? Is it acceptable for her to cry over this?

_____

_____

19:46–
22:15

2. Why is Bob crying? Is it acceptable for him to cry over this?

_____

_____

AFTER YOU WATCH

LANGUAGE POINT: NOUN CLAUSES AS OBJECTS

A *noun clause* is used in the same way as a *noun*. It can be the *subject* or the *object of the verb*. In the following sentences from the video, the noun clauses, are all *objects of the verb*. Unscramble the sentences.

Identify the subject (**S**) , the main verb (**V**) and the object (**O**). Then rewrite the sentence in appropriate word order.

EXAMPLE :   Judy   what she wanted   knew
                 S   **O**   **V**

Judy knew what she wanted.

1. thought   Judy   that he would be open and generous

_____

2. Jo-Ann   to know   wanted   how he felt

_____

3. where they were going to live   Judy   wanted   to find out

_____

4. was Judy's depression   typical of married women   felt   Jo-Ann

_____

5. Bob   what his role   needed   to know   was

_____

VOCABULARY CHECK: COLLOQUIAL EXPRESSIONS

The following excerpts are from the video. What do the *italicized* words and expressions mean? Circle the correct answers.

1. I immediately said, "Now, that's a lady I can *relate to*."
   a. understand and   c. have a family with
      communicate with
   b. meet the relatives of   d. talk about

2. She said exactly what she wanted . . . and I said, "*I can handle that*."
   a. I can lift that by the handle.   c. I can manage that.
   b. I can see what you mean.   d. I can help with that.

3. And it *lets everybody off the hook* in terms of taking the responsibility for figuring out their own life goals.
   a. allows everybody to go fishing   c. allows nobody to go fishing
   b. removes everybody's obligation   d. removes nobody's obligation

4. But even though Judy had finally *stood her ground*, three months had gone by and Bob still hadn't made a decision.
   a. maintained her position   c. planted a tree in her garden
   b. stood on the floor   d. walked alone on her land

5. Bob has to *come to terms with* his emotions himself.
   a. destroy
   b. face
   c. go to school with
   d. hide

DISCUSSION

Work in groups. Discuss your answers to the following questions.

1. Do you know a couple who is having trouble in their marriage or relationship? How could therapy help them?
2. If therapy is too expensive or culturally unacceptable in a situation, what other forms of help can a person seek?
3. In your opinion, do Judy and Bob have a better chance at a happy marriage now? Explain your answer.

ROLE PLAY

Work in groups of three. One student will play the role of counselor. Another will play the role of the wife. The third will play the role of the husband. Read the situation and the role descriptions below and decide who will play each role. After a ten-minute preparation, begin the interview.

THE SITUATION: **A Marriage Counseling Session**

A couple whose marriage is in trouble has arranged to see a marriage counselor. After the first session, the counselor has determined that the root of their problems is jealousy.

ROLE DESCRIPTION: **Marriage Counselor**

You want to help the couple to listen to and understand each other better. You often paraphrase what each one says in order to clarify meaning. You use phrases like "so what you mean is . . ." and, "in other words." You ask questions beginning with *What, Where, When, Who,* or *Why* to clarify statements they make, but you avoid stating your own opinions in your questions and comments.

ROLE DESCRIPTION: **Wife**

You have been married for fifteen years. At the beginning of the marriage, you worked, but since the arrival of three children, you have dedicated yourself to your family. You feel left out of the career world and have recently become jealous of your husband's secretary. You think he is spending an unusual amount of time with her.

ROLE DESCRIPTION: **Husband**

You have worked your way up the ladder to become a successful executive. You are proud of your family and your numerous accomplishments, but you find it increasingly difficult to communicate with your wife, who seems more involved with the children's accomplishments than your own. Your work gives you the opportunity to travel, and you take advantage of this because it is a way to escape from your unhappy home life.

READING

Read the article below to find out about a technique one married couple used to to save their marriage. Then answer the questions that follow.

THE POWER OF COMMUNICATION

Harry and Sally were best friends in elementary school, and she was barely seventeen when he proposed to her. They got married a year later. By the time Sally was twenty, she had a year-old child and was eight months pregnant with another. Harry worked as a mechanic at a nearby gasoline station and took additional jobs on weekends to help make ends meet. Life wasn't easy. Harry had little time to pay attention to his family. There was always another customer with a car to be fixed "right away." Sally was in an endless cycle of changing diapers, cooking, and cleaning. They soon realized that their marriage had become a nightmare. They were headed for disaster until they started to play their "secret game."

One night, as Sally washed dishes after supper, she broke into tears. Harry asked her softly what was wrong, but she refused to tell him. She knew he was doing his best to provide for the family. If she told him how much she hated being inside the house all day, constantly running after the baby and trying to make a decent meal with the few ingredients they could afford, it would hurt his feelings and make him feel worthless. Harry gently persuaded her to go outside and sit next to him on the porch. As they swayed back and forth in an old swing, he reminded her of the "secret game" they used to play as children.

It was a guessing game that allowed them to fantasize about what they wanted to be when they grew up. The rules were simple. One person would think of an occupation and give hints about what it was. For every "wrong guess" the person giving the hints wrote the first letter of the occupation. The game wasn't tied to occupations alone. Sometimes they used it to describe feelings they had or things they wanted. As kids, they had soon forgotten the rules of the game and quickly began to describe what they would do to obtain their goals.

Now, as adults, they found they were very much the same. Sally was able to communicate to Harry her wish to work part time as a secretary to get a break from the children. In time she wanted to go back to school to become a legal secretary. Harry was able to admit that he still had his childhood dream of becoming a mechanical engineer. Every evening for the next two weeks they discussed ways to achieve these goals. Life became fun again. Hope and excitement replaced despair and frustration. Harry avoided taking additional weekend work and started to spend more time figuring out how he and Sally could achieve their goals.

For six years, they struggled — making endless arrangements for childcare, working, and studying. However, there was one major difference from their old life. They now combined their efforts to please each other, to provide for their children's needs, and to build a better future for themselves and the children. Harry and Sally rediscovered the power of communication, which helped keep their love and their marriage alive.

1. Describe Harry's daily routine in the early years of their marriage.

   _____

2. Describe Sally's routine.

   _____

3. Why didn't Sally want to tell Harry what was wrong?

   _____

4. How did Harry find out what Sally really wanted to do?

   _____

5. What did each of them want to do?

   _____

6. What saved this marriage?

   _____

WRITING

Complete one of the following activities.

1. As a professional therapist, you have been asked to give a speech that includes the most common issues of concern in marriages. Write a list of the most important issues.
2. You have been arguing with your husband or wife about a certain issue. Write a letter to him or her. In the first paragraph, state his or her side of the issue. In the second paragraph, state your side of the issue and the reasons for your feelings. Conclude the letter by asking him or her to discuss this further. Suggest a time, date, and place where you can meet and discuss the issue until you reach an understanding.

Segment 3
Children of Divorce

From: *20/20*, 3/4/82
Begin: 25:09
Length: 15:06

BEFORE YOU WATCH

TALKING POINTS

Work in groups. Discuss your answers to the following questions.

1. Is divorce common in the country in which you were born?
   Explain your answer.

2. What are some ways that divorce affects children? Make a list.

3. What are some things that parents who are planning to divorce can
   do to prepare their children?

PREDICTING

The video is about children of divorced parents. Write your answers to the
following questions. Then compare your answers with those of another
student.

1. What do you already know about children of divorced parents?

   _____

   _____

2. What are you unsure of about children of divorced parents?

   _____

   _____

3. What do you expect to learn about children of divorced parents?

_____

_____

KEY WORDS

The words and phrases below are used on the video. Put each word or phrase into one of the categories on the chart. Then compare your chart with that of another student.

afraid	embarrassed	misunderstood
aggressive	frightened	non-custodial parent
bewildered	ignored	sad
child support	issues	settlement
court	mad	single parent
custody	mediator	visitations

WORDS THAT DESCRIBE FEELINGS	TERMS USED IN DIVORCE
_____	_____
_____	_____
_____	_____
_____	_____
_____	_____
_____	_____
_____	_____
_____	_____

GETTING THE MAIN IDEA

Watch the news report and listen for the answers to the following questions. Take brief notes on the answers. Then compare your answers with those of another student.

25:18–
40:24

> **Who** is doing **what**?
> **Why** is this important?
> **How** are these studies different from other studies, and **how many** children are affected?

Who?	
What?	
Why?	
How . . . different?	
How many?	

CHECKING YOUR PREDICTIONS

Look at your answer to question number 3 in the PREDICTING exercise on pages 21 and 22. Watch the video again. Did you learn what you expected to learn from the video?

25:18–
40:24

WHAT'S MISSING?

Listen to Hugh Downs' introduction to the news report. Fill in the missing words.

25:18–
26:17

Hugh Downs: These children are part of one of the fastest-growing

(1)_____ groups in the United States—the children of

(2)_____ , thirteen million of them in this country right now.

And (3)_____ , because divorcing adults are so often

(4)_____ with their own pain, the children's voices are

(5)_____ heard. But that's changing—now there is new

attention to the children, new studies on the (6)_____ of

divorce from their point of view. And here with a report is Bob Brown.

TRUE OR FALSE?

26:41–
28:10

Watch the next part of the video. Are the following statements *true* or *false*? Write **T** (true) or **F** (false). Make the false sentences true by changing one or two words.

1. _____ Michael is three years old.

2. _____ Michael's parents are already divorced.

3. _____ Michael is afraid of being separated from his parents.

4. _____ The way Michael plays reflects (shows) how he feels.

5. _____ Michael wants his own family to stay together.

6. _____ Michael sees divorce as a kind of death in the family.

LISTENING FOR DETAILS

28:10–
29:14

Watch the video. Circle the correct answer.

1. By how much has the divorce rate in the United States increased over the last twenty years?
   a. By 2.5 percent.
   b. By 25 percent.
   c. By 250 percent.

2. How many children below the age of eighteen have divorced parents?
   a. Thirty thousand.
   b. Thirteen million.
   c. Thirty million.

3. Who co-authored a study on divorced children that was published in 1980?
   a. Joan Kelly.
   b. Hugh Downs.
   c. Bob Brown.

4. How many children are affected by new divorces each year?
   a. One million.
   b. Two million.
   c. Three million.

5. What does Dr. Joan Kelly feel is "sort of embarrassing"?
   a. That so many people are divorced.
   b. That millions of children have been affected by divorce.
   c. That mental health professionals didn't think enough about the issues of divorce.

NOTETAKING

Watch the next part of the video and take brief notes on the answers to the following questions. Then compare your notes with those of another student.

29:15–
38:18

1. Who wrote *The Kids' Book of Divorce?* Why was it written?

   _____

2. The 1927 film *Children of Divorce* reflects who the children of divorce were and what they might become. Explain this view.

   _____

3. How are more recent films such as *Shoot the Moon* different from the older film *Children of Divorce?*

   _____

4. What is one of the most important recent findings that research has made in studying divorce?

   _____

5. What is the typical response of pre-school children to a divorce? nine- to twelve-year-old children?

   _____

6. How do divorcing parents often misuse their children who are between the ages of thirteen and eighteen?

   _____

7. How do adolescent children of divorcing parents view their parents?

   _____

MAKING TRUE SENTENCES

Watch the video. Then use the chart to make four true sentences. Write the sentences at the top of page 26.

32:30–
40:03

Most children wish . . . How well children do after a divorce . . . We don't know . . . It is difficult for adults . . .	1. is determined by the contact they have with both parents. 2. to be in touch with their children's feelings. 3. their parents would reunite. 4. whether children of divorce are more likely to have a divorce when they grow up.

1. _____

2. _____

3. _____

4. _____

LANGUAGE POINT: PARTICIPIAL ADJECTIVES

In the video, Hugh Downs uses participial forms of the verbs *divorce* and *absorb* as adjectives when he says, "And, sadly, *divorcing* adults are so often *absorbed* with their own pain, the children's voices are seldom heard." Complete the following sentences by using the appropriate participial form of the verb in parentheses.

1. Dr. Kelly thinks it is (embarrass) *embarrassing* that people in the mental health profession didn't think through the issues of divorce.

2. The little girl in the 1927 film *Children of Divorce* was (frighten) _____ when her mother left her.

3. Divorce can be a (shatter) _____ experience for children.

4. Preschool children are often (bewilder) _____ by the absence of a parent.

5. The absence of a parent is very (frighten) _____ to a very young child.

6. Children around the ages of five or six are very likely to become (depress) _____ by the absence of a parent.

VOCABULARY CHECK: COLLOQUIAL EXPRESSIONS

The following sentences are from the video. What do the *italicized* expressions mean? Circle the answer that is closest in meaning.

1. We're going to see and hear children — who often *couldn't care less about* the issues adults are familiar with.
   a. have no interest in   c. think a lot about
   b. pay a lot of money for   d. can't understand

2. And that's one of the most important findings researchers have made — identifying the ways in which children *deal with* divorce.
   a. manage or cope with   c. cause
   b. give reasons for   d. do business with

3. . . . so you can *get on with* this business of developing who you are.
   a. forget about
   b. avoid
   c. end
   d. continue

4. She used to call me *her* rock.
   a. the person who depended on her the most
   b. the person who hated her the most
   c. the person she depended on the most
   d. the person who liked her the most

5. It can be a *real drag* to wake up in the morning and come downstairs to find somebody standing in your living room in a bathrobe, [somebody] that you've never seen before.
   a. an exciting game
   b. a big disappointment
   c. a wonderful surprise
   d. a truly happy moment

6. Maybe his children won't have to *go through* it.
   a. enjoy
   b. suffer or experience
   c. move away from
   d. remember

DISCUSSION

Work in groups. Discuss your answers to the following questions.

1. In your opinion, should children in divorce cases be consulted about their parents' visitation rights? If so, how old should the children be before their ideas are considered as part of the divorce settlement?

2. If a child's parents decide not to honor the final divorce settlement, should there be a law to force the parents to adhere to the terms that have been agreed upon to protect the child's rights? Give examples of how this might work in terms of (a) visitation rights, (b) child support, (c) medical expenses, and (d) educational expenses.

3. Do you believe that children of divorce are more likely to get divorced themselves when they grow up?

ROLE PLAY

Work in groups of three. One student will play the role of the father. Another student will play the role of the mother. The third student will play the role of the child. After a ten-minute preparation, begin the role play.

THE SITUATION: **A Disagreement over Visitation Rights**

Divorced parents have agreed to share parental responsibilities, although their nine-year-old child will be living primarily with the mother. The parents cannot agree on how often the father should see the child.

ROLE DESCRIPTION: **Father**

You are a successful psychologist who has flexible working hours. Your patients come to your office early in the morning or late in the evening after work. You would like to see your child every afternoon when he/she comes home from school. You feel that you shouldn't have to wait until weekends in order to see your child.

ROLE DESCRIPTION: **Mother**

You are a successful business executive who believes that children need discipline in order to succeed. You would like your child to come home from school every day and finish his/her homework so that when you come home from work you can spend quality time together. You feel the father would disrupt that discipline since he is not as strict as you are.

ROLE DESCRIPTION: **Child**

You would like to see both of your parents as much as possible. Create a plan that will accommodate everyone's schedule

READING

Rose is a newspaper columnist who answers letters from people with personal problems. Read Rose's newspaper column and then answer the questions that follow.

Rose answers your problems ...

Dear Rose:

My husband and I had been married for sixteen years, eight of which were miserable. I tried to stay in the marriage as long as I could for the sake of the children, but the constant arguments and my husband's heavy drinking became too much to tolerate. After an agonizing decision-making period, I decided to file for divorce. The decision was based on the welfare of the children as well as my own. I didn't want them to grow up with his kind of behavior as a model.

What I thought would take a short time to settle has now taken eight months. The process of divorce has turned into a nightmare. The arguments, which are now mostly over money and property, have increased. But what hurts the most is seeing how the children are caught in the middle of this adult warfare and are being used by their father to get back at me.

He often speaks to me rudely in front of the kids, knowing that I will not make an attempt to protest because of their presence. He demands that they show him affection when I'm present, knowing that they will yield to his demands to avoid another fight between us. Sometimes he motions them to stay away from me simply to make me feel unwanted.

This situation is affecting the children tremendously. There is hardly any code of discipline to follow in the home since there is always a dispute as to which adult is in charge. The children have become unruly and uncommunicative. My son, who was a good student until now, is failing one of his courses at school, and my daughter, who has always been open and social with people, has become quiet and withdrawn.

My lawyer says it will be several more months before the divorce is settled. In the meantime, I'm supposed to endure this situation. What can I do?

Worried Mom

Dear Worried Mom,

Divorce is never an easy issue, especially when children are involved. However, you have made your choice and now you must help your children get through it.

Children caught in the middle of a divorce often need reassurance that despite the differences between Mom and Dad, both of their parents love them. Children often assume that their parents will always stay together. When that sense of continuity is shaken, they fear what will happen to them and who will take care of them. Children need to know that issues pertaining to their health, education, and shelter will be made with their best interests at heart.

It is very important to talk to your children and try to reduce their level of stress. While it is not necessary to discuss every intricate detail of the divorce with them, it is good to reassure them that divorce is a temporary process that makes everyone edgy, including the adults, but that it is a progressive process that will eventually come to an end.

The most important thing you can do for your children is to tell them over and over that the divorce is between the two parents who are having difficulties with each other. It is not a divorce between parents and children. Reassure them that both parents still love them.

Rose

1. How long has the woman been married?

_____

2. Why did the woman file for divorce?

_____

3. What are the woman and her husband arguing about?

_____

4. How are the children being treated unfairly?

_____

5. How are the woman's children being affected by the divorce?

_____

6. According to Rose, what things do the children need to be reassured of?

_____

7. What do children often assume about their parents?

_____

8. What is the most important thing this woman can do for her children?

_____

WRITING

Complete one of the following activities.

1. Imagine that you have a friend who is going through a divorce. Write a letter to the friend advising him or her how to handle things so that the children will suffer the minimum amount of pain.

2. Write a list of at least three different visitation arrangements that a non-custodial parent could choose from. Describe each arrangement in two or three sentences.

Segment 4
Old Problems: New Cures

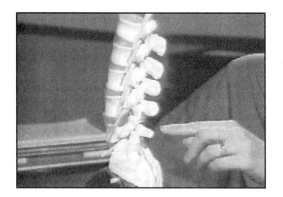

From: *20/20* 12/9/82
Begin: 40:25
Length: 12:28

*Portions of this video may be too graphic for some students. Viewer discretion is advised.

BEFORE YOU WATCH

TALKING POINTS

Work in groups. Discuss your answers to the following questions.

1. Have you, or has anyone you know, had back surgery? If so, what were the results?
2. What are some of the remedies for back pain?
3. If you suddenly experienced a backache, what would you do? Where would you go for help?

PREDICTING

Work in groups. The news report is about new cures for backache. Write down five questions you think will be answered on the video.

1. _____

2. _____

3. _____

4. _____

5. _____

KEY WORDS

The following words are used in the video. Which words do you think you will hear together? Match the words on the left with those in the box. The first one has been done for you.

1. muscle   _spasm_

2. bed   _____

3. slipped   _____

4. spinal   _____

5. sciatic   _____

6. bladder   _____

7. medical   _____

8. surgical   _____

9. CAT   _____

10. local   _____

cord
emergency
~~spasm~~
control
rest
disc
anesthetic
nerve
procedure
scan

WHILE YOU WATCH

40:33–
52:53

GETTING THE MAIN IDEA

Watch the news report and listen for the answers to the following questions. Take brief notes on the answers. Then compare your answers with those of another student.

Who does backache affect?

What are the various symptoms and
What are the alternative cures?

Why are there alternative cures?

Who?	
What . . . symptoms?	
What . . . cures?	
Why?	

CHECKING YOUR PREDICTIONS

Look at the questions you wrote in the PREDICTING exercise on page 31. Watch the video again. Are any of your questions answered on the video? What answers are given?

40:33–
52:53

WHAT'S MISSING?

Listen to Hugh Downs' introduction to the news report. Fill in the missing words.

41:20–
42:00

Hugh Downs: Backache — it affects everyone, from laborers to office workers to weekend gardeners. And as old and (1) _____ a problem as it is, (2) _____ is (3) _____ about its causes or its (4) _____. The major problem in back research is that different (5) _____ within the back can create the (6)_____ pain, and that's what makes (7) _____ so (8)_____ . At the bioengineering laboratories at the University of Illinois, some of the mysteries of the back are now being explored. For example, we know that (9) _____ lifting injures many backs, but we don't know why. This experiment shows how holding a ten-pound weight at arm's length rather than close to the body (10) _____ the back muscles by fifteen times.

TRUE OR FALSE?

Watch the video. Are the following statements *true* or *false*? Write **T** (true) or **F** (false). Make the false statements true by changing one or two words.

42:01–
45:10

1. _____ Eighty percent of back pain is caused only by muscle spasm.

2. _____ Muscle spasm can sometimes be alleviated (lessened) by bed rest.

3. _____ A slipped disc is the bulging center of a spinal disc pressing against the nerves.

4. _____ A discectomy is not the most common surgical procedure on backs.

5. _____ Most herniated (slipped) discs heal themselves, even without surgery.

6. _____ The United States has the lowest rate of back surgery in the world.

7. _____ Over 30% of the back operations in the United States fail.

8. _____ Leaders in the field of back surgery encourage surgery as a cure.

9. _____ Dr. Augustus White thinks a patient should always get a second opinion when surgery is suggested for the first time.

10. _____ The most common cause of failed back surgery is choosing the wrong doctor.

CHECKING WHAT YOU SEE

42:36–
43:38

Watch the video with the *sound off*. Check (✓) the items that you see on the video.

1. ❑ spine   4. ❑ nucleus   7. ❑ lungs

2. ❑ intestines   5. ❑ heart   8. ❑ stomach

3. ❑ spinal disc 6. ❑ annulus   9. ❑ vertebrae

NOTETAKING

45:11–
45:40

I. Watch the next part of the video. Hugh Downs mentions four symptoms that indicate the right candidate for back surgery. List the four symptoms.

1. _____

2. _____

3. _____

4. _____

45:41–
45:58

II. On the next part of the video, Dr. Belkin lists ten risks that patients face from surgery. Watch the video as many times as necessary and complete the following list of risks that are mentioned by Dr. Belkin.

RISKS OF BACK SURGERY	
1. failure to relieve pain	6. _____
2. _____	7. _____
3. loss of bowel control	8. spinal fluid leak
4. _____	9. clots in leg
5. loss of motor control in the lower extremities	10. _____

INFORMATION MATCH

Watch the next part of the video. Match the names with the correct information.

45:59–
50:34

a. Abby Summersgill	c. Beth Israel	e. Judith Walker
b. John Dunleavy	d. Stuart Belkin	f. Carol Warfield

1. _____ is a doctor who is studying the use of an enzyme called chymnopapain for the treatment of back pain.

2. _____ chose chymnopapain treatment for himself when he had a slipped disc.

3. _____ has had four operations since he hurt his back on his job as a milkman.

4. _____ is a Boston hospital that offers alternative treatments for back pain.

5. _____ had chronic burning pain in his legs from an inflamed nerve root.

6. _____ is an anesthesiologist who favors the injection of a steroid in cases where the nerve root is inflamed.

7. _____ received treatment that involved the injection of steroids and the use of a special type of X-ray called fluoroscopy.

8. _____ is an anesthesiologist and neurophysiologist whose research shows that electrical stimulation of certain nerves can relieve back pain.

LISTENING FOR DETAILS

Watch the last part of the video. Circle the correct answers.

50:35–
52:53

1. Which of the following is true about electrical stimulation treatment for back pain?
   a. The success rate is higher in patients who have had surgery.
   b. The success rate is lower in patients who have had surgery.
   c. The success rate is the same for both groups of patients.

2. What were the results of John Dunleavy's steroid treatment?
   a. There was no change in his condition.
   b. His condition improved, and he felt well enough to swim.
   c. His condition got worse, and he decided that surgery was the only answer to his pain.

3. What were the results of Abby Summersgill's steroid treatment?
   a. There was no change in her condition.
   b. She became more uncomfortable and felt more tension in her back.
   c. She was able to go water-skiing two weeks after the injection.

4. Which of the following is true?
   a. If you have back pain, you should always avoid surgery.
   b. If you are not the right candidate for back surgery, you should examine the alternatives.
   c. Hugh Downs' back surgery was unsuccessful.

AFTER YOU WATCH

LANGUAGE POINT: GIVING ADVICE

On the video Dr. Augustus White gives the following advice:

> "We say that if the surgery is suggested for the first time, you certainly always *ought to* have two opinions."

> Here are four common ways of giving advice:
>
>   a. You ought to . . .
>
>   b. You should . . .
>
>   c. You shouldn't . . .
>
>   d. You'd better . . .

Use expression *a, b, c,* or *d* as indicated to complete each of the following dialogues with your advice. The first one has been done for you.

1. A: I've had a backache for a week. What should I do?*(b)*.
   B: *You should* consider a back pain specialist _____

2. A: My doctor has recommended back surgery. What should I do? *(a)*
   B: _____

3. A: This piece of furniture is very heavy, but it needs to be moved. How should I move it? *(d)*
   B: _____

4. A: The pain killers I'm taking are very expensive, and they're really not effective. *(c)*
   B: _____

5. A: I don't want to have surgery on my back. What should I do to relieve the pain? *(b)*
   B: _____

VOCABULARY CHECK

On the video, the word *back* is used in the phrases below. Write a short definition of each phrase. The first one has been done for you.

1. back exercises: *movements that exercise the muscles of the back*
2. back pain:_____
3. back muscles: _____
4. back sufferers: _____
5. back surgery: _____
6. back surgeon: _____

DISCUSSION

Work in groups. Discuss your answers to the following questions.

1. Both patients on the video benefited from experimental treatments using steroids. Yet, steroids and other experimental drugs may cause uncomfortable side effects. Do you know of any side effects related to the use of experimental drugs? If so, describe them. Would you chose to be treated with these drugs?

2. It is hard to imagine that patients as young as those on the video could be crippled by back pain. Yet, this is the reality that some backache victims experience. Do you know of anyone who has suffered backache problems? If so, describe what you know of this person's experience with back pain.

READING

Read the following case study about one woman's experience with back pain and the choices she faced to relieve her problem. Then answer the questions that follow.

CONNIE'S DILEMMA

Two weeks had passed since the incident occurred. It hadn't seemed important at the time. Connie Cabetas had done nothing out of the ordinary — only her usual housework chores. With three kids in the household, bending down to pick up clothes and toys from the floor was an everyday occurrence. Moving furniture to clean the obscure corners that her seven-month-old baby could crawl into was a must and had become part of the routine. As she was shoving the nine-foot long sofa back into its place with her knee, however, she felt something snap in her back. At first, it didn't hurt. But as the day wore on, the intense pain that began at the bottom of her spine traveled down her right leg all the way to the middle of her calf and became intensifying and crippling.

When her husband, Tom, came home that evening, he found her immobile in bed. Dinner had not been made, and the kids were unkempt. Her condition did not improve during the next few days, and her pain became excruciating. Over-the-counter medication was no help. Her mother volunteered to come over to help with the kids, but this temporary arrangement was short-lived. In a matter of days, Connie's well-kept home had become a wreck. Her family needed her desperately, but she could hardly drag herself out of bed for more that a few minutes at a time. She was finally able to get an emergency appointment with a well-respected orthopedic surgeon. The doctor diagnosed her as having sciatica, a condition caused by a herniated disc pressing against the sciatic nerve. The doctor described the following alternatives to Connie and Tom. There was no guarantee that any of them would permanently solve her condition, but it was all that the medical field could suggest.

One alternative was back surgery. Statistics show that 30 percent of back surgeries in the United States have been unsuccessful. These failures include the possibility of increased pain, loss of bowel and bladder control, loss of motor control in the lower extremities, loss of sensation, infection, spinal fluid leak, clots in the legs, and clots in lungs. On the other hand, roughly 70 percent of all patients have had some success with surgery and were able to lead fairly normal lives after four to six weeks of recuperation. Another fact that the Cabetas had to consider was medical insurance coverage. Back surgery was well covered under their medical plan.

The second alternative was an exercise and rehabilitation plan. Experimental physical and psychological therapy shows a 40 percent success rate among patients with back problems. The intense therapy would require Connie to live in a clinic for three to four weeks. Moreover, since the therapy was still in its experimental stages, it would not be covered by their health insurance. The cost would affect their finances substantially.

The third alternative was nothing more than to continue bed rest for as long as needed — in addition to taking pain killers that could lead to addiction. Still, time might give Connie's body the chance to rehabilitate itself. Although the drugs were expensive, they were covered by the family's medical plan.

The last option was to take an experimental drug injected into the epidural space at the base of the spine. The drug has a 60 percent success rate. The 40 percent failure rate includes side effects that are not entirely predictable by the medical community. Since it was an experimental drug, the procedure was not covered by their medical insurance.

Connie and Tom had to make a decision, and it needed to be made in a hurry. Their family life was crumbling down around them.

1. How long has Connie been suffering from this condition?

_____

2. What incident led to her back problem?

_____

3. What kind of doctor did she go to see?

_____

4. What was the doctor's diagnosis? What causes this condition?

_____

5. What were the four choices Connie had?

_____

6. Which option would you have advised them to take?

_____

ROLE PLAY

Work in groups of three. One student will play the role of the surgeon. Another will play the role of Connie. The third will play the role of Connie's husband, Tom. Read the situation and the role descriptions below and decide who will play each role. After a ten-minute preparation, begin the discussion.

SITUATION: **A Discussion with a Surgeon**

A housewife named Connie Cabetas has been afflicted by a severe backache that has crippled her. Her husband and doctor are advising her to take different options to alleviate her condition.

ROLE DESCRIPTION: **Tom**

You are a middle management executive in a large company, and you plan to continue working with the company. Although your family medical benefits are adequate for average medical expenses, you are not equipped to handle a big medical debt at the moment. You are very concerned about the cost of your wife's treatment and urge her to select an option that is covered by your insurance.

ROLE DESCRIPTION: **The Surgeon**

You have suggested four alternatives that are available to Connie. Your expertise is back surgery. If Connie chooses any of the other options, she will not come to you for treatment.

ROLE DESCRIPTION: **Connie**

Confused and in pain, you want to make the best choice for yourself. You are inclined to take physical therapy treatments, even though this would increase your family's financial burden.

WRITING

Complete one of the following activities.

1. Using the information that you have learned from the video, write a letter to a friend to discourage her from having back surgery. Name other alternatives that she might pursue.

2. The video, the role play, and the reading all allude to the fact that there are alternative methods to surgery. Write up a list of all the steps you would take as a patient to explore all the available possibilities before you would choose surgery.

Segment 5
Dr. Jonas Salk, Discoverer of Polio Vaccine

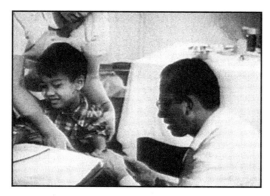

From: *World News Tonight,* 6/14/91
Begin: 52:55
Length: 4:18

BEFORE YOU WATCH

TALKING POINTS

Work in groups. Discuss your answers to the following questions.

1. What is a vaccine? How does a vaccine help prevent infection?

2. Have you ever been vaccinated? Against what diseases?

3. Name some diseases that children are vaccinated against. How often do children need to be vaccinated against these diseases?

4. Where can you go to be vaccinated? Do you have to pay?

PREDICTING

Work in groups. Based on the title of the news report, *Dr. Jonas Salk, Discover of Polio Vaccine,* write down five questions you think will be answered on the video.

1. _____

2. _____

3. _____

4. _____

5. _____

KEY WORDS

The *italicized* words in the sentences below will help you understand the video. Study the sentences. Then write your own definition of each word.

1. Dr. Jonas Salk is an international *celebrity* in the field of medicine.

   celebrity:_____

2. If people are *infected* with the HIV virus, they may develop AIDS.

   infected:_____

3. Medical *researchers* have been looking for a cure for AIDS.

   researchers:_____

4. When one *approach* doesn't work, it is necessary to try another.

   approach:_____

5. One day we hope to be able to *immunize* people against AIDS.

   immunize:_____

6. A weakness in the immune system is one of the *symptoms* of AIDS.

   symptoms: _____

7. More than one scientist has proposed a *hypothesis* about the origin of AIDS.

   hypothesis: _____

8. Many years ago, Jonas Salk received a *grant* to develop a polio vaccine.

   grant:_____

WHILE YOU WATCH

53:03–
57:21

GETTING THE MAIN IDEA

Watch the news report and listen for the answers to the following questions. Take brief notes on the answers. Then compare your answers with those of another student.

Who is doing **what? Where?**

Why and **how** are they doing this?

Who?	
What?	
Where?	
Why?	
How?	

CHECKING YOUR PREDICTIONS

Look at the questions you wrote in the PREDICTING exercise on pages 41. Watch the video. Which of your questions are answered on the video? What answers are given?

53:03–
57:21

WHAT'S MISSING?

Listen again to Peter Jennings' introduction to the news report. Fill in the missing words.

53:03–
53:25

Peter Jennings: Finally this evening, our Person of the Week. There are younger members of our (1)_____ who will not know him. There are middle-aged Americans and older who will (2)_____ him as an international (3)_____ of their youth, which (4)_____ us how long it has been since he did so much to (5)_____ the world of a dreadful (6)_____. We chose him this week because he has (7)_____ and is still (8)_____ others to find the way.

TRUE OR FALSE?

Watch the next part of the news report again. Are the following statements *true* or *false*? Write **T** (true) or **F** (false). Make the false sentences true by changing one or two words.

53:26–
54:22

1. _____ Immunization normally has been used in individuals who are not yet infected in order to prevent the establishment of infection.

2. _____ In 1986, it occurred to Dr. Jonas Salk that AIDS researchers were coming closer to finding a cure for AIDS.

3. _____ Dr. Jonas Salk began to think of another approach to cure the AIDS virus.

4. _____ His new approach would immunize individuals who are already infected before the development of symptoms of disease.

5. _____ When a person first becomes infected with the AIDS virus, the body's immune system collapses immediately.

6. _____ Dr. Salk's approach would give the immune system a boost that would help keep the disease from spreading.

WHO SAYS WHAT?

Watch the next part of the video and listen to the comments made by Dr. Allan Goldstein and Dr. Anthony Fauci. Who says what? Check (✓) the appropriate boxes.

54:23–55:11

Who said . . . ?	Dr. Allan Goldstein	Dr. Anthony Fauci
1. "He challenged conventional wisdom, and he developed a new idea. He had a hunch, and he stuck to it."		
2. "It may not necessarily be safe. It may very well be that later on you're going to see negative effects."		
3. "Right now, in June of 1991, we haven't seen efficacy in this study. We've seen some promising, intriguing results that deserve further study."		

CHECKING WHAT YOU SEE

Watch the video with the *sound off*. What do you see? Check (✓) the appropriate boxes.

55:23–56:40

1. ❑ a doctor looking at test tubes
2. ❑ nurses bathing patients
3. ❑ children with polio
4. ❑ x-ray machines

5. ❑ breathing machines
6. ❑ children being injected
7. ❑ patients using walkers
8. ❑ a monkey being injected

PUTTING EVENTS IN ORDER

Read the sentences below. Then watch the video and put the events in the correct order. Number them 1 to 8. The first event has been numbered for you.

55:23–
56:40

____ Salk went into research at the University of Michigan.

____ Salk's polio vaccine was declared safe and effective.

____ Salk and his medical team discovered a vaccine against influenza.

____ Salk won a grant to do research on a polio vaccine.

____ The Salk polio vaccine was replaced by the Sabin vaccine.

_1_ Salk received his M.D. from New York University.

____ Salk discovered the polio vaccine.

____ Thousands of children were stricken with polio.

MAKING INFERENCES

Watch the video. Put a check (✓) next to each sentence that represents what Dr. Salk meant when he said:

56:41–
56:51

"It means that we're not putting all our eggs in one basket, and as we know how nature works, it's through diversity. Soon nature will give us the answer."

1. ____ We're not placing all our hopes on one method.

2. ____ Nature works in a linear and simple way.

3. ____ Nature works in many different ways.

4. ____ The answer for this disease will come from nature.

5. ____ We need to stop looking and give nature a chance to do its work.

CHECKING WHAT YOU HEAR

Watch the video. Check (✓) the statements that agree with what you hear.

1. ____ Jonas Salk is 86 years old now.

2. ____ Jonas Salk is nearly ready to retire.

3. ____ Jonas Salk believes he is onto something in his fight against AIDS.

4. ____ Jonas Salk believes his experiment must go on.

56:52–
57:21

5. ____ Jonas Salk is very sick and soon will not be able to do anymore research.

LANGUAGE POINT: PRESENT PERFECT PROGRESSIVE VS. SIMPLE PAST TENSE

The *simple past tense* is used to describe an event that was completed at a specified time in the past.

> **Example:** Dr. Salk *received* his M.D. in 1939.
> (He received his degree at a specified time: in 1939.)

The *present perfect progressive tense* is used to describe an event that began in the past, continues in the present, and may continue in the future.

> **Example:** Dr. Salk *has been searching* for results most of his life.
> (His search continues in the present time.)

Complete the following sentences with the simple past or the present perfect progressive tense of the verb in parentheses, whichever is more appropriate. The first one has been done for you.

1. For more than a decade now, researchers (look) *have been looking* for ways to fight the AIDS virus.

2. In 1986, Dr. Salk (begin) _____ to think about a new approach.

3. Dr. Salk (win) _____ a grant to do research on the polio vaccine in 1948.

4. In 1955, researchers (say) _____ the polio vaccine as safe and effective.

5. Since 1986, Dr. Salk (experiment) _____ with a vaccine approach against the AIDS virus.

6. Since Dr. Salk started to experiment with a vaccine approach, other AIDS researchers (watch) _____ the test results very closely.

VOCABULARY CHECK

The *italicized* words are used on the video. Cross out the word or phrase that *does not* have a similar meaning to the word in *italics*. The first one has been done for you.

1. *persevered*	~~gave up~~	continued	persisted
2. *ultimately*	finally	at last	right away
3. *conventional*	unusual	traditional	customary
4. *wisdom*	intelligence	insight	ignorance
5. *hunch*	idea	feeling	problem
6. *inhibiting*	encouraging	limiting	restricting

7. *optimism*	hopefulness	confidence	fear
8. *efficacy*	validity	effectiveness	weakness
9. *intriguing*	unattractive	fascinating	interesting
10. *diversity*	variety	similarity	difference

IDIOMATIC EXPRESSIONS

The excerpts below are from the video. What do the *italicized* expressions mean? Circle the correct answer.

1. . . . AIDS researchers were becoming *bogged down* trying to find drugs that would cure or prevent AIDS.
   a. excited about
   b. unable to make progress
   c. starting to get good results

2. But over time, the body's immune system *gives in*.
   a. collapses
   b. improves
   c. gets bigger

3. . . . and *keep* the disease *in check* using the AIDS virus itself.
   a. control
   b. promote
   c. ignore

4. He had a hunch, and he *stuck to* it.
   a. continued with
   b. got stopped by
   c. forgot about

5. . . . In the fight against AIDS, he does believe he *is onto something*.
   a. has become one of the most important researchers of all time
   b. has some evidence that could lead to an important discovery
   c. has overcome all the difficulties possible in his research

WORD FORMS

The sentences below are from the video. The *italicized* words are either nouns, verbs, or adjectives. Fill in the chart that follows by adding the other forms of each word. The first one has been done for you. In some categories, there is more than one possibility. When you finish, compare your chart with that of another student.

1. But over time, the body's *immune* system gives in.
2. Salk began to *experiment* with a vaccine.
3. He *challenged* conventional wisdom, and he *developed* a new idea.

4. The announcement this week that another scientific team had taken his idea . . . and was testing it with some *success* has been met with *optimism* and caution.
5. A high school teacher of his called him a *perfectionist*.
6. And so we choose Jonas Salk, ever the *controversial* figure.

NOUN	VERB	ADJECTIVE
immunity, immunization	*immunize*	immune
	experiment	
	challenged	
	developed	
success		
optimism		
perfectionist		
		controversial

DISCUSSION

Work in groups. Discuss your answers to the following questions.

1. What did you learn from the video?
2. AIDS has infected and ultimately taken many lives throughout the world. Do you know any victims? In what ways has the virus spread?
3. As researchers continue to search for a cure for AIDS, we have been warned to take measures to ensure our own health. Name some ways that you can avoid the risk of being infected with AIDS.

ROLE PLAY

Work in groups of three. One student will play the role of the interviewer. The other two students will play the roles of the grant candidates. Read the situation and the role descriptions below and decide who will play which roles. After a ten-minute preparation, begin the interview.

THE SITUATION: **A Screening Interview**

The Research Foundation for the Humanities is sponsoring an award to help find the best way to stop the spread of AIDS. The winning candidate will receive a two-million-dollar grant to launch an aggressive campaign against the spread of AIDS and will become internationally known in the field of AIDS prevention. Many psychologists, medical doctors, business people, and politicians have applied for the grant, but only one person can receive it. Two candidates are being interviewed. Both are highly talented, successful, and well-known medical researchers.

ROLE DESCRIPTION: **Interviewer**

You are an interviewer for the Research Foundation for the Humanities. Your job is to screen candidates so that the one with the best qualifications and the best ideas. The winner will have the responsibility of trying to stop a world epidemic. You need to be very careful in making your choice.

ROLE DESCRIPTION: **Candidate A**

You are a well-known research scientist who wants this grant because it will allow you to play a vital role in medical research. You feel that your background qualifies you for the job. In addition to your scientific qualifications, a member of your family has died from AIDS. As a result of this tragedy you feel great compassion for the victims of this disease.

ROLE DESCRIPTION: **Candidate B**

You are a well-known scientist who knows that if you get this grant, you and your family will be financially secure. It will also bring you international fame and open many doors for the future. You are personally disgusted with the AIDS disease, but you want to take on the job for the sake of its prestige and material benefits.

READING

How much do you know about AIDS? Take this test and find out. The answers follow the test.

THE AIDS TEST

1. Which of the following is false? You can get AIDS by . . .
   a. having unprotected sex.
   b. hugging a person who has AIDS.
   c sharing an injection needle.
   d. getting an infected blood transfusion.
   e. being born of an infected mother.

2. You can tell that a person has AIDS by . . .
   a. the way the person looks.
   b. how thin the person is.
   c. the results of an HIV antibody test.
   d. how much the person eats.
   e. how often the person catches cold.

3. Is the following statement true or false?
   HIV is the same thing as AIDS.

4. Is the following statement true or false?
   Scientists have not yet found a cure for AIDS.

Answers to **THE AIDS TEST**:

1. (b) is *false*. In order for you to get AIDS, the HIV virus must work its way into your blood stream. This can happen through the other four suggested answers. You cannot get the HIV virus into your bloodstream by hugging anyone.

2. (c) is *true*. You cannot tell if a person is infected simply by looks. It often takes weeks, or months, even years before a person who has been infected by the HIV virus shows any AIDS symptoms.

3. The statement is *false*. HIV is an abbreviation for "Human Immunodeficiency Virus," a virus that affects the immune system in a person's body and reduces resistance to illness. AIDS stands for "Acquired Immunodeficiency Syndrome," which is a combination of physical signs and symptoms that occur because your immune system has been damaged by the HIV virus.

4. The statement is *true*. Scientists have NOT yet found a cure for this deadly disease.

WRITING

Complete one of the following activities

1. You have a chance to write a letter to an AIDS victim. Imagine that you know this person from school or work. You want to give courage to this individual because you know that soon he or she will face death.

2. You want to warn teenagers of the dangers that AIDS may have on their young lives, but you know that it is very difficult to find a good attention-getter. Write a riddle or a rap that transmits a warning about the dangers of AIDS.

Segment 6
Ninetysomething

From: *Prime Time Live,* 9/3/92
Begin: 57:22
Length: 13:12

BEFORE YOU WATCH

TALKING POINTS

Work in groups. Discuss your answers to the following questions.

1. Describe "old age." How do people feel? How do they look? What do they do?
2. When should men and women retire from their regular jobs? Should elderly people work at all after they retire?
3. If you live to be ninety years old or older, how do you think you will be spending your time?

PREDICTING

The video is about new discoveries that scientists are making about aging. What do you think you will see and hear on the video? Write down three items under each of the headings below. Then compare your answers with those of another student.

SIGHTS	WORDS
(things you expect to see)	(words you expect to hear)

1. _____   _____
2. _____   _____
3. _____   _____

KEY WORDS

The *italicized* words will help you understand the video. Study the definitions. Then use each word or phrase in a sentence of your own.

1. *myth:* a false story or idea that many people may believe is true

   _____

2. *gerontology center:* a laboratory that does research related to old age

   _____

3. *brain cells:* the smallest independent parts of the brain

   _____

4. *breakthrough:* a major discovery in science, medicine, technology, etc.

   _____

5. *symptoms:* signs of illness in a person's body, such as fever, swelling, or pain

   _____

6. *sex drive:* the natural need or desire for sexual intercourse

   _____

7. *Alzheimer's:* a disease occurring in older people that includes loss of memory

   _____

8. *stroke*: a sudden and severe illness that affects the brain, often killing people or causing one side of the body to become paralyzed

   _____

9. *life span*: the total length of time that someone is likely to live

   _____

10. *chronologic age*: the actual number of years a person has been alive (also called "chronological age")

   _____

WHILE YOU WATCH

GETTING THE MAIN IDEA

Watch the news report and listen for the answers to the following questions. Take brief notes on the answers. Then compare your answers with those of another student.

0:57:29–
1:10:41

> **What** are scientists studying?
>
> **What** have they discovered?
>
> **When** do they expect to make a breakthrough?
>
> **Why** is this research important?

What . . . studying?	
What . . . discovered?	
When?	
Why?	

CHECKING YOUR PREDICTIONS

Look at the lists you made in the PREDICTING exercise on pages 51. Watch the video and check (✓) the items that you actually see and hear.

0:57:29–
1:10:41

IDENTIFYING WHAT YOU SEE

Watch the first part of the video with the *sound off*. What do you see on the video? Match the time codes with the phrases on the right.

58:15–
59:18

1. 58:18-58:36 _____   a. an elderly woman walking
2. 58:37-58:56 _____   b. an elderly potter working in a studio
3. 58:57-59:03 _____   c. an elderly man training for a race
4. 59:04-59:18 ___   d. an elderly woman playing piano in a night club

WHO'S WHO?

Look at the chart below. Then watch the video with the *sound on* and check (✓) the appropriate boxes.

58:21–
59:38

Who . . .?	Sadie Colar	Beatrice Wood	Paul Spangler
1. is training for the New York Marathon			
2. is 92 years old			
3. plays piano in New Orleans night clubs and hotels?			
4. gets up at 6:00 A.M. to go shopping			
5. works eight hours a day as a potter			
6. turned 98 years old the year this news story was broadcast			

WHAT'S MISSING?

Listen to Diane Sawyer as she talks about some new discoveries that scientists are making. Fill in the missing words.

0:59:40–
1:00:08 **Diane Sawyer:** Beatrice, Sadie, Paul — they may seem like fortune's

(1)_____ , blessed in their old age. But here's a

(2)_____ for you. Scientists are now convinced that in the

next two (3)_____ , the 90s are going to be like that for a

whole lot of us. They're going to feel like the back end of

(4)_____ age, because every month in (5)_____

around the country, scientists are making startling discoveries,

breaking open the old (6)_____ about the aging

(7)_____ and the (8)_____ .

TRUE OR FALSE?

Watch the video. Are the following statements *true* or *false*? Write **T** (true) or **F** (false). Make the false statements true by changing one or two words.

1:00:09–
1:01:33

1 _____ Dr. Ed Schneider works at the Andrus Gerontology Center.

2. _____ The Andrus Gerontology Center is studying different chemicals that keep the brain vital.

3. _____ Dr. Ed Schneider thinks that the process of losing brain cells as we age probably can't be stopped.

4. _____ When scientists at U.S.C. added "nerve growth factor" to damaged brain cells of rats, more of the brain cells died.

5. _____ Scientists have already started similar tests on human brain cells.

6. _____ Ten years from now, people may be taking pills to change their brain chemistry.

NOTETAKING

Watch the video and take brief notes on the answers to the following questions. Then compare your notes with those of another student.

1:01:34–
1:04:12

1. What is Deprenyl?

_____

2. How did Dr. Morton Shulman feel after he first took Deprenyl?

_____

3. What brain disease does Dr. Shulman have?

_____

4. What are some of the positive effects of Deprenyl, according to people who have used it?

_____

5. Is there any proof that Deprenyl does anything for healthy humans?

_____

INFORMATION MATCH

Watch the next part of the video. Match the names with the correct information.

1:03:58–
1:09:46

    a. Beatrice Wood   b. Sadie Colar   c. Paul Spangles
    d. Maria Fiatarone   e. Roy Walford

1. _____ is a retired doctor.

2. _____ says her mind brings her back to life when she's exhausted.

3. _____ can hear a new song and then play it by ear.

4. _____ produces 50 or 60 pieces of pottery a year.

5. _____ started playing the piano on riverboats at the age of 20.

6. _____ has trouble remembering new songs.

7. _____ is conducting strength tests of people in their 80s and 90s.

8. _____ does diet research using mice equal to 90-year-old humans.

9. _____ is convinced the human life span can be extended to 150.

10. _____ is in training for the New York Marathon.

AFTER YOU WATCH

LANGUAGE POINT: STATING A POSSIBILITY

On the video, Dr. Ed Schneider states a possibility when he says, "there *may* be a pill that will allow us to have the same brain power at age 90 . . . that we had at age 20." The modals *may* and *might* are often used to state a possibility. Rewrite the following sentences, using *may* or *might*. The first one has been done for you.

1. There is a possibility that we are wrong about the process of aging (*might*)

    We might be wrong about the process of aging.
_____

2. Maybe Deprenyl would work for healthy humans. (*may*)

_____

3. With new techniques, it is possible that doctors will be able to repair damaged brain cells. (*might*)

_____

4. There is a possibility that calorie restriction is related to longer life span. (*may*)

_____

5. Maybe the old myth that aging can't be stopped is wrong. (*might*)

_____

6. It is possible that some people feel and act young simply because of their spirit. (*may*)

_____

VOCABULARY CHECK

The following sentences are from the video. What do the *italicized* words and expressions mean? Circle the answer that is closest in meaning.

1. After mile eight, he's *a little out of breath*.
   a. breathing rather quickly
   b. breathing very slowly
   c. breathing normally
   d. not able to breathe at all

2. During the week, Sadie Colar plays two or three *sets* a night in New Orleans clubs and hotels.
   a. songs
   b. groups of songs
   c. piano concertos
   d. parts of a song

3. And in this studio a *prolific* potter works eight hours a day on pieces that sell for up to $30,000.
   a. elderly
   b. wealthy
   c. very lazy
   d. very productive

4. Dr. Ed Schneider is the head of the Andrus Gerontology Center at U.S.C., one of a dozen places *feverishly* working with different chemicals to keep the brain vital.
   a. slowly and without much interest
   b. very quickly and excitedly
   c. carelessly
   d. carefully

5. Doctors will tell you if you're healthy, to take a pill *on a gamble* is crazy.
   a. as a chance at something
      better
   b. that costs a lot of money
   c. when you are far away
   d. as an escape

6. Sadie Colar isn't waiting for science to *come up with* a pill.
   a. find or produce
   b. stop work on
   c. advertise
   d. sell

7. The work still keeps her brain *supple*, especially her memory.
   a. unable to respond to ideas
   b. quick to respond to ideas
   c. soft and easy to damage
   d. tired and hard to use

8. Sadie is *living proof* that, contrary to what everybody thinks, memory doesn't have to collapse in old age.
   a. a live example of a fact
   b. a live example of a myth
   c. a dead example of a theory
   d. a false example of a theory

9. What she learned, even about people who've had strokes and heart attacks, *defies* common logic.
   a. improves or increases
   b. makes worse or decreases
   c. challenges or disagrees with
   d. proves or agrees with

10. What age would you like to be *in the hereafter*?
    a. for the rest of your life
    b. at this moment
    c. in the places around here
    d. in life after death

DISCUSSION

Work in groups. Discuss your answers to the following questions.

1. Do you personally know any people in their 90s who are as active and healthy as Sadie Colar, Paul Spangler, and Beatrice Wood? If so, what do you think is the "secret" of their long and healthy lives?

2. As parents get older, there may be a role reversal between them and their children. Sometimes children end up taking care of their elderly parents. Is this true in the country in which you were born? Explain your answer.

3. Do you do anything special to increase your own chances of having a long and healthy life? Explain your answer.

ROLE PLAY

Work in pairs. One student will play the role of the high school student. The other student will play the role of a grandparent. Read the situation and the role descriptions that follow and decide who will play each role. After a ten-minute preparation, begin the interview.

An Interview for a School Project

Century High is having a social science fair. Students have been asked to interview one of their grandparents to find out what life used to be like for them. The students will ask the grandparents about the hardships they have experienced and how they were able to overcome them, how science and technology have affected their lives, and what life is like for them now.

ROLE DESCRIPTION: **High School Student**

You want to feel proud of yourself and your family by presenting unusual facts that make your grandfather or grandmother unique. Make up a list of questions you will use to interview your grandparent.

ROLE DESCRIPTION: **Grandparent**

You want to give your grandchild a sense of worth because of what you have been able to accomplish in life and how this is reflected in your present lifestyle. Be prepared to answer your grandchild's questions about what life used to be like for you and what life is like for you now. To do this effectively you need to think about what you would like your lifestyle to be when you became elderly.

READING

Read the article below to find out about Tina and Esther, two elderly sisters with very different lifestyles. Then answer the questions that follow.

SISTERS

It is mistakenly assumed that siblings born and raised in the same family have more similarities than differences in their personalities. In fact, the differences develop and grow throughout their lives and show themselves especially as siblings approach their golden years. Below is a description of the daily routines of two sisters. Both lead active lives in their advanced ages. Yet their lifestyles are very different.

Tina wakes up early in the morning and makes breakfast for herself and her husband. As he reads the newspaper after breakfast, she hurries around the house, tidying up the bedroom, bathroom, and living room. By mid-morning she is usually in the kitchen planning or preparing the evening meal. Her husband usually busies himself gardening or by doing small repairs around the house. At noon both sit down for a quick lunch. Afterward, he finishes reading the paper while Tina tries to catch up on her sewing. At two in the afternoon, their youngest grandson comes home from elementary school. Tina is sure to have a snack ready for him. She then makes herself available to help him with his schoolwork. An hour or so later, the second grandson comes home from high school. He too is greeted with a snack and a smile. It is grandpa's job to make sure that this young man gets his studies done. Around five, when the grandchildren leave with their mother, Tina serves an early dinner for herself and her husband. After dinner they settle down to a night of TV or a rented video.

Esther, on the other hand, lives alone with her grandson, who is now twenty-one. Her day starts around five o'clock in the morning. She gets dressed, feeds the cats, eats breakfast, and is off for a brisk walk. By the time she returns, her grandson is getting ready to go to school. She gives him a hand with breakfast. As soon as he is out the door, she gets on the phone with one of her girlfriends and begins to plan the day or to confirm previous plans. Household chores are done by 10:00 A.M. so that she can enjoy the rest of the day. Her grandson is left a plate of food, which he just needs to reheat at night. Sometimes she goes bargain shopping or perhaps strolls by the many cafes along the beach. There are many activities for senior citizens. These often include luncheons, which she enjoys. Afternoons are reserved for political activities. She works as a volunteer for various fund-raising efforts for local and national figures. This often leads to evening engagements, which she also enjoys.

Both sisters feel they are enjoying life to the fullest at their age—Tina through family-related activities, and Esther through more social and politically related ones. Despite differences in lifestyles, these two women continue to lead active and fulfilling lives.

1. What are the two main ideas in the first paragraph?

_____

2. What is Tina's lifestyle like?

_____

3. What is Esther's lifestyle like?

_____

4. What are some of the differences between their daily routines?

_____

5. What are some of the similarities in their daily routines?

_____

6. Which sister maintains a busier lifestyle?

_____

WRITING

Complete one of the following activities.

1. Scientists have finally made the breakthrough we have all been waiting for. Soon there will be a pill on the market that will allow your brain to remain as active in your 90s as it was in your 20s. Create a one-page advertisement that will help market this pill. Include a picture to represent the product and write a caption of no more than five lines.

2. You are an active retiree who has been taking the wonder pill that is now available to repair damaged brain cells. You lead a very active life, physically and socially. Fill in your diary for a typical week. Write what you plan to do each day.

Mon.	Tues.	Wed.	Thurs.	Fri.	Sat.	Sun.

Segment 7

Ability Grouping Can Stifle a Child's Educational Development

From: *American Agenda*, 9/5/90
Begin: 1:10:42
Length: 4:47

BEFORE YOU WATCH

TALKING POINTS

Work in groups. Discuss your answers to the following questions.

1. What do you think the terms "tracking" and "ability grouping" mean in relation to students?

2. Does the "ability grouping" system exist in schools in the country in which you were born? If so, describe.

3. Do you think separating high-achievers from low-achievers in the classroom is helpful in the education of children? helpful to the teacher? Give reasons for your answers.

PREDICTING

Work in groups. Based on the title of the news report and the picture above, predict the kinds of information you think will be included on the video.

1. _____

2. _____

3. _____

4. _____

5. _____

KEY WORDS

Work in groups. Look at the words and phrases below. Check (✓) the words and phrases you think you will hear on the video and explain why.

- ❑ racism
- ❑ alternatives
- ❑ sorting
- ❑ discrepancy
- ❑ test scores
- ❑ equal opportunity

- ❑ assumptions
- ❑ ability grouping
- ❑ development
- ❑ self-esteem
- ❑ held back
- ❑ develop a scar

WHILE YOU WATCH

GETTING THE MAIN IDEA

Watch the news report and listen for the answers to the following questions. Take brief notes on the answers. Then compare your answers with those of another student.

1:10:50–
1:15:37

Who is doing **what, where**?

Why and **how** are they doing this?

Who?	
What?	
Where?	
Why?	
How?	

CHECKING YOUR PREDICTIONS

Look at your answers to the PREDICTING exercise on page 61 and above. Watch the video again and check (✓) the kinds of information that are actually included on the video.

1:10:50–
1:15:37

LISTENING FOR DETAILS
Watch the video. Circle the correct answers.

1:10:50–
1:13:06

1. What is "tracking" or "ability grouping"?
   a. Putting kids together who have similar abilities.
   b. Putting kids together who have the same interests.
   c. Putting kids together who have different abilities.

2. Which statement is false?
   a. Ability grouping benefits children.
   b. Ability grouping doesn't work.
   c. Ability grouping can affect a child's total educational development.

3. Is it possible that what a child ends up being as an adult is partly determined by an educational sorting process?
   a. No, it isn't.
   b. Yes, it is.
   c. The video doesn't say.

4. To which students did teachers in the past usually target their lessons?
   a. The brighter kids.
   b. The slower kids.
   c. The kids who were in the middle.

5. How was ability grouping supposed to work in theory?
   a. It would keep fast kids from being held back and slow kids from being frustrated.
   b It would help fast kids get good jobs and dumb kids get bad jobs.
   c. It would allow the fast kids to help the slow kids.

6. What really happens to kids who are in "tracks"?
   a. The test kids score of the slow kids improve significantly.
   b. Both the fast kids' and slow kids' test scores drop over the years.
   c. The test scores of the fast kids increase steadily.

WHAT'S MISSING?
Listen to Bill Blakemore's introduction to the news report. Fill in the missing words.

1:11:17–
1:11:41

Bill Blakemore: Is it (1)_____ that whether you become a

successful (2)_____ or a prison (3)_____, a

research (4)_____ or an unskilled (5)_____, a

professional (6)_____ or one of the chronically

(7)_____, is it possible that might be (8)_____

partly by an educational sorting process which begins when

you're six years old? Yes, quite possible!

TRUE OR FALSE ?

1:11:17–
1:13:12

Watch the video. Are the following statements *true* or *false*? Write **T** (true) or **F** (false). Make the false statements true by changing one or two words.

1. _____ An educational sorting process can determine what you are going to be as an adult.

2. _____ "Tracking" is uncommon in most American schools.

3. _____ "Tracking" is used only in high schools.

4. _____ Most parents in the United States accept tracking decisions made by educators.

5. _____ The theory behind "tracking" has been found not to work.

6. _____ The graph shown on the video proves that children at every level of ability in a track system tend to achieve less.

NOTETAKING

1:12:11–
1:13:47

Watch the video. Three reasons for using "tracking" are given, along with three reasons why "tracking" doesn't work. Write each reason next to the name of the speaker who states it.

WHY "TRACKING" IS USED:

Woman	If you mix the groups . . .
Bill Blakemore	The theory is that . . .
Robert Berry	We're putting them there . . .

WHY "TRACKING" DOESN'T WORK:

Anne Wheelock	But they don't see the deficits . . .
Bill Blakemore	Children at every . . .
Jeannie Oakes	If you're told that . . .

IDENTIFYING PEOPLE'S FEELINGS

Two teenagers on the video give their feelings about "ability grouping" by making three different statements. What feelings are the teenagers trying to express? Match the words and phrases in the box with the time codes.

1:13:13–
1:14:18

dumb	not worth the effort	labeled
limited	little teacher attention	poor teacher assessment

1. 01:13:13 – 01:13:25 _____ _____

2. 01:13:25 – 01:13:31 _____ _____

3. 01:14:06 – 01:14:18 _____ _____

CHECKING WHAT YOU HEAR

Watch the next part of the news report. What problems connected with "tracking" are mentioned? Check (✓) the appropriate boxes.

1:14:19–
1:14:39

1. ❑ less qualified teachers
2. ❑ less access to learning resources
3. ❑ smaller schools
4. ❑ less stimulating curriculum
5. ❑ less time for lunch
6. ❑ less attention

AFTER YOU WATCH

LANGUAGE POINT: EXPRESSING CONDITIONS AND RESULTS

On the video, a woman expresses a condition and a result when she says, "If you mix the groups totally, you're going to lose not only the bright child or the slow child, you're going to lose both." If-clauses are used to express a condition. The condition is then followed by a result. If-clauses may appear at the beginning or at the end of a sentence. When an if-clause appears at the beginning of a sentence, it is followed by a comma.

Use an if-clause to combine each pair of sentences into one. On the first line, write the if-clause at the beginning of the sentence. On the second line, write the if-clause at the end of the sentence. The first one has been done for you.

1. Children are motivated. They become better learners.
   If children are motivated, they become better learners.
   Children become better learners if they are motivated.

2. Fast learners help slow learners. Both types of children will benefit.

   _____

   _____

3. Children are put in either fast or slow tracks. Their test results drop.

   _____

   _____

4. Children are placed in a slow class. They feel the teacher doesn't care.

   _____

   _____

5. Children are given a stimulating curriculum. They are willing to learn more.

   _____

   _____

VOCABULARY CHECK

The *italicized* words are used in the video. Cross out the word or phrase in each row that *does not* have a similar meaning to the *italicized* word. The first one has been done for you.

1. *successful* — flourishing — prosperous — ~~decreasing~~
2. *sorted* — put together — classified — separated
3. *theory* — philosophy — action — idea
4. *disheartened* — demoralized — discouraged — happy
5. *benefit* — good — disadvantage — gain
6. *logical* — reasonable — sensible — crazy
7. *research* — investigation — guess — study
8. *deficit* — excess — lack — insufficiency
9. *accumulate* — collect — gather — disperse
10. *self-esteem* — self-respect — shame — pride

DISCUSSION

Work in groups. Discuss your answers to the following questions.

1. "Tracking" may prevent students from taking courses they really want simply because the courses are not available on their track schedules. If you were in this situation, how would you feel? What would you do?

2. Business corporations often use a "track system." For example, employees who are not in a managerial track may find it impossible to get high-ranking jobs. Have you ever been in this situation, or do you know another person who has been? If so, give examples.

3. In some countries "social status" follows a track system. For example, if a family enjoys a high economic status, certain benefits may be available to them that are not available to the rest of the people. Is this true in the country in which you were born? If so, describe.

4. From what you know about the United States, do you think social status in the United States follows a "track system"? Why or why not?

ROLE PLAY

Work in groups of four. One student will play the role of the parent, the second the role of the student, the third the role of the teacher, and the fourth the role of the principal. Read the situation and the role descriptions below and decide who will play each role. After a ten-minute preparation, begin the conference.

THE SITUATION: **A Parent-Teacher Conference**

A high school student wants to take an electronics course but has been told by the electronics teacher that he cannot take it because his math grade is too low. The student has complained to his parent. Now the teacher, parent, and student are meeting with the school principal to discuss the situation.

ROLE DESCRIPTION: **Teacher**

According to school policy, the student is not qualified to take the course. You do not want to make exceptions to the rule. First, it might get you in trouble with administration; second, you might end up with a low-ability student who would make teaching the class more difficult.

ROLE DESCRIPTION: **Principal**

You support the teacher's decision because it is the practical thing to do. You are looking at the issue not from a learning point of view but from a scheduling point of view. You would never admit this to a parent, however. You know that if you make an exception now, you will have to make other exceptions later, and if you change the student's schedule to include the electronics course, all his other classes will have to be changed, too. All this will increase your paperwork.

ROLE DESCRIPTION: **Parent**

Although your child's math grade is low, you feel he has the ability to do well in the electronics course. He has a knack for fixing things around the house and is highly motivated to take the course. You think that if he does well in the course, it might give him the incentive to do even better in his other courses.

ROLE DESCRIPTION: **High School Student**

You want to take the electronics course because you like the subject and know you will do well in it. You know your math grade is low simply because you hated math and did not put any effort into it. You have done well in other courses and think it is unfair to look only at your math grade. You suspect that issues other than your math grade are involved.

READING

How do students feel about the ability grouping system? The following questionnaire was distributed to a group of high school students in Florida. The number in parentheses show the percentages of total students who gave that answer on the questionnaire. Read the questionnaire. Then answer the questions that follow.

QUESTIONNAIRE

1. Have you participated in any special "tracking" or "ability group" programs in your educational experience?
   a. Yes (94%) b. No (0%) c. Don't know (6%)

2. How would you rate the track or group to which you were assigned?
   a. Fast (66%) b. Slow (2%) c. In the middle (32%)

3. How did you feel about being placed in that track or group?
   a. Good (78%) b. Bad (3%) c. Not sure (19%)

---

3. How did you feel about being placed in that track or group?
   a. Good (78%)   b. Bad (3%)   c. Not sure (19%)

4. Have you ever felt that you wanted to be in a different track or with a different group of students?
   a. Yes (43%)   b. No (44%)   c. Not sure (13%)

5. If you had been put in a different group, how do you think your grades would have been?
   a. Higher (40%)   b. Lower (10%)   c. About the same (50%)

6. Have you ever been denied the opportunity to take a course you wanted because it wasn't offered at a time your track schedule allowed?
   a. Yes (34%)   b. No (54%)   c. Not sure (12%)

7. Do you feel that you are in a track program now?
   a. Yes (88%)   b. No (8%)   c. Not sure (4%)

8. Are you happy with your present track placement?
   a. Yes (78%)   b. No (7%)   c. Not sure (15%)

9. Of the courses offered at your school, can you take any course you want?
   a. Yes (46%)   b. No (34%)   c. Don't know (20%)

10. Who would you go to in order to get this information?
    a. Teacher (6%)   c. Parent (2%)
    b. Counselor (76%)   d. Don't know (16%)

1. According to the statistics obtained from the questionnaire, to which kind of track (fast, middle, or slow) did most of the students feel they had been assigned?

   _____

2. Read the statistics for question number 3, then circle the statement that best describes the students' responses.
   a. Most of the students were dissatisfied with being placed in the track to which they were assigned.
   b. Most of the students were satisfied with being placed in the track to which they were assigned.

c. Most of the students were happy that they were getting out earlier from school.

d. Most of the students weren't sure whether they liked being in the track to which they were assigned.

3. Explain the difference between question 3 and question 8.

_____

_____

4. According to the results of the questionnaire, which of the following statements *does not* describe the general experiences or feelings of the students?

a. Most of the students are happy to be in the track to which they have been assigned.

b. Most of the students know where to go for help if they have a question about courses offered at their school.

c. Most of the students have been denied access to particular courses because of their track assignment.

WRITING

Complete one of the following activities.

1. You are a high school principal who is considering introducing the ability grouping system into your school. Using information from the video, make up two lists. In one list include all the good points associated with tracking. In the other list include all the bad points associated with tracking.

2. You are a parent of a high school student whose school is planning to introduce a tracking system. Write a letter to the principal of your child's school. In the letter, explain why you think the tracking system is a good idea or a bad idea.

Segment 8

Cooperative Learning: A Good Alternative to Ability Grouping

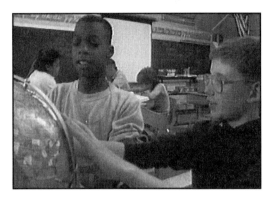

From: *American Agenda*, 9/6/90
Begin: 1:15:38
Length: 4:38

BEFORE YOU WATCH

TALKING POINTS

1. "Cooperative learning" is based on the idea that students will learn more by working together in pairs or in small groups than they will by working alone. Do you agree? Why or why not?

2. When different groups of students are working together, the classroom can become quite noisy. Do you think that this noise interferes with learning? Describe your most successful type of learning environment.

3. Another idea behind the idea of "cooperative learning" is that each student in the class is a potential teacher. It is not necessary for students to wait for just one teacher to give them the right answer. They have a variety of other teachers available. Do you think this sharing of power in the classroom is good educational practice? Explain your answer.

PREDICTING

Work in groups. Based on the title of the news report and the picture above, predict the kinds of information you think will be included on the video.

1. _____

2. _____

3. _____

4. _____

5. _____

KEY WORDS

The following words are used on the video. Which words do you think you will hear together? Match the words on the left with those in the box.

1. growing _____
2. remarkable _____
3. trouble _____
4. cooperative _____
5. test _____
6. spending _____
7. ability _____
8. each _____
9. figure _____
10. one _____

free
other
numbers
results
time
to one
things out
scores
learning
groups

WHILE YOU WATCH

GETTING THE MAIN IDEA

1:15:45–
1:20:13

Watch the news report and listen for the answers to the following questions. Take brief notes on the answers. Then compare your answers with those of another student.

Who is doing **what, where?**

Why and **how** are they doing this?

Who?	
What?	
Where?	
Why?	
How?	

CHECKING YOUR PREDICTIONS

Look at your answers to the predicting exercise on page 72. Watch the video again and check (✓) the kinds of information that are actually included on the video.

1:15:45– 1:20:13

WHAT'S MISSING?

Listen to Bill Blakemore's introduction to the news report. Fill in the missing words.

1:16:16– 1:16:41

Bill Blakemore: In a (1)_____ number of schools, teachers are asking the kids to put their (2) _____ together, and it's getting remarkable (3)_____. It's called (4) "_____ learning," and teachers say it's producing (5) _____ test (6)_____— even for slow students, trouble-free school (7)_____, fewer discipline problems. And it can do all this without (8) _____ children into fixed ability groups.

LISTENING FOR DETAILS

Watch the video. Circle the correct answers.

1:16:16– 1:19:14

1. What are teachers asking the kids to do?
   a. Sit around in groups and solve problems in class.
   b. Do more homework that will help their grades go up.
   c. Sit quietly in rows and copy from the board.

2. What do teachers who believe in "cooperative learning" say about test scores?
   a. Kids are getting lower test scores than before.
   b. Kids are getting higher test scores than before.
   c. Test scores remain the same.

3. Who sets the task for the students?
   a. The students themselves.
   b. The textbook.
   c. The teacher.

4. A male teacher says, "I can zero in one-to-one or one-to-four." What does this lead us to understand?
   a. He will give little attention to that group.
   b. He prefers to work with a larger group.
   c. He can give students in small groups more individual attention.

5. A female teacher says, "If I'm here teaching and somebody gets stuck,they have someone there." Why does this teacher believe this is good?
   a. She doesn't have to teach as much.
   b. Students don't have to wait for her to give the answer.
   c. Students become dependent on one another.

6. Which of the following *does not* reflect the theory behind "cooperative learning"?
   a. It helps kids to find out from each other how to get answers.
   b. It allows teachers to relax.
   c. It allows slow kids to work at their own pace.

7. Which of the following is an aspect of "cooperative learning"?
   a. It separates students of different levels of accomplishment.
   b. It mixes together students of different levels of accomplishment
   c. It successfully hides the problems of the slower students.

8. In which way is "cooperative learning" also an individualized system?
   a. Each student gets points, not for the most right answers, but for individual improvement.
   b. Each teacher keeps a separate folder for each of her students.
   c. Each group competes against other groups in the classroom.

CHECKING WHAT YOU SEE

Watch the video with the *sound off*. Check (✓) the behaviors that you see.

1:16:45–
1:18:14

1. ❑ students sitting in rows
2. ❑ Students helping one another
3. ❑ ethnically and racially mixed groups working together
4. ❑ students fooling around and not paying attention
5. ❑ students engaged in figuring out problems
6. ❑ only elementary school kids involved in groups
7. ❑ kids working on the floor
8. ❑ girls working only with girls

CHECKING WHAT YOU HEAR

Watch the video and check (✓) the benefits of "cooperative learning" that are mentioned.

1:17:00–
1:17:41

1. ❑ The teacher doesn't have to teach any more.
2. ❑ Kids are good at re-explaining and individualizing material.
3. ❑ A kid can always ask a friend to explain something.
4. ❑ Communication is improved.
5. ❑ The teacher can work with larger groups.

6. ❑ The kids can continue working when the teacher is with another group.
7. ❑ The teacher can take a break any time she wants to.
8. ❑ The kids have a roomful of teachers.

TRUE OF FALSE?

Watch the next part of the video. Are the following statements *true* or *false*? Write **T** (true) or **F** (false). Make the false statements true by changing one or two words.

1:17:42–1:18:46

1. _____ Kids can move at their own pace.

2. _____ A kid is forced to move at the same pace as the group.

3. _____ If no one in the group knows the answer, a student can go to a another group to get help.

4. _____ Kids are not allowed to look at each others' papers.

5. _____ Kids learn to share answers and discuss problems.

6. _____ When kids teach other kids, they learn the material better themselves.

7. _____ When you teach someone something, you forget it quickly.

8. _____ Learning facts from the board is still the best way.

NOTETAKING

Watch the last part of the video and take brief notes on the answers to the following questions. Then compare your notes with those of another student.

1:19:15–1:20:13

1. How are the students tested?

_____

2. How do teachers in the most successful programs keep records?

_____

3. When would a teacher encourage a student to work alone?

_____

4. How do teachers rearrange the learning groups every few weeks?

_____

5. What two feelings of accomplishment do kids get from "cooperative learning"?

_____

_____

AFTER YOU WATCH

LANGUAGE POINT: EXPRESSING NECESSITY AND LACK OF NECESSITY

On the video a teacher expresses lack of necessity when she says, "If I'm here teaching and somebody gets stuck, they have someone there, they don't have to wait." In "cooperative learning," students and teachers have a new set of classroom rules to follow. What things do they have to do? What things don't they have to do? Complete each of the following sentences with *have to* or *don't have to* and one of the verbs in the box. The first one has been done for you.

wait	depend
move on	communicate
set	~~figure out~~
spend	do

1. Students ___*have to figure out*___ problems in their groups.
2. Students _____ all their time sitting in rows.
3. Teachers _____ the task for students.
4. Students who need help _____ for the teacher.
5. Students _____ with each other.
6. Teachers _____ most of the talking.
7. Students _____ until they understand the material.
8. Students _____ only on the teacher.

WORD FORMS

The sentences below are from the video. The *italicized* words are either nouns, verbs, or adjectives. Fill in the chart on page 77 by adding the other forms of each word. The first one has been done for you. In some categories, there is more than one possibility. When you finish, compare your chart with that of another student.

1. It's called "*cooperative* learning," and *teachers* say it's producing higher test scores.
2. The teachers can improve their *communication* too, moving from group to group . . . and do so while all kids stay *engaged*.
3. I think the greatest *benefit* is the kids can move at their own pace.

76   UNIT 3: APPROACHING EDUCATION IN NEW WAYS

4. Having kids *explain* things to each other uses what teachers have always known: you really get to know something when you teach it.
5. They feel much better about themselves because they're more *accepted* by the other children in the classroom.
6. They match kids whose strengths *compliment* each other.
7. I think kids in cooperative education get a feeling of "I know how to get *information*."

NOUN	VERB	ADJECTIVE
		cooperative
teachers		
communication	*communicate*	*communicative*
		engaged
benefit		
	explain	
		accepted
	compliment	
information		

DISCUSSION

Work in groups. Discuss your answers to the following questions.

1. Have you ever worked in small groups in the classroom? If so, which method do you think works better for you: working in small groups or individually? Explain.
2. Compare cooperative learning to the teaching methods used in the country in which you were born. Explain the similarities and differences.
3. In your opinion, do some subjects lend themselves better to "cooperative learning" than others? Are there some subjects that would be better taught in the traditional way? Explain your answer.
4. Think of situations outside the classroom when you have worked in a small group or with someone else to solve a problem. How did you feel? Can that same feeling be transferred to solving problems in the classroom? Give details.

ROLE PLAY

Work in groups of three. One student will play the role of the teacher. The other two will each play the role of one of the students. Read the situation and the role description below and decide who will play each role. After a ten-minute preparation, begin the role play.

THE SITUATION: **Solving a Conflict Within a Group**

A teacher has given a group ten minutes to come up with an idea for a commercial for a new medical center that's about to open. Two of the students in the group are stubborn about their ideas and do not want to compromise. The teacher tries to help the students work out the conflict before time runs out.

ROLE SITUATION: **Teacher**

You know that two of the students in this group are very bright and will therefore challenge each other. You realize, however, that they, like the rest of the students in the class, need to learn how to work with one another and not argue all the time. You have put them together specifically for that reason. You are not going to make any exceptions to the time limit. If the group does not come to a mutual agreement, they will fail the task.

ROLE SITUATION: **Student A**

You are convinced that the most successful ad will be one that appeals to people's feelings of compassion. You think service should be identified as the primary selling point. You envision an ad showing caring nurses taking care of the elderly and young children. You realize that Student B has different ideas, but you are too proud to even consider listening to them. You have always been an "A" student in this class, and you know that the teacher will think your idea is a good one.

ROLE SITUATION: **Student B**

You have been interested in the medical field for some time and know a lot about the latest medical equipment and procedures. You feel that this information, combined with the idea of a limited hospital stay, will most appeal to people. You feel that your peers should give you credit for being able to provide this type of information, which is unfamiliar to them.

READING

In the video, you heard several teachers give their opinions about "cooperative learning." How do other teachers feel about this method? A group of elementary school teachers who have been using "cooperative learning" for at least five years were asked the following questions. Some of the teachers' responses are listed after each of the questions. Read the questions and the teachers' responses. Then answer the questions that follow.

SURVEY: COOPERATIVE LEARNING

For the purpose of this survey, "cooperative learning" is defined as **"The practice of forming small heterogeneous groups, that is, kids with different learning abilities, that work together for the purpose of completing a task that has been assigned by the teacher."**

1. What factors do you take into consideration when grouping students?
   a. *"I look at academic ability, social behavior, and equal distribution of the sexes."*
   b. *"I try to look at the student's "working personality": How will they relate to others in the group? Can they put any differences aside so that they can complete a task at hand? Do they know how to compromise, offer ideas, and make decisions?"*
   c. *"I think about learning styles."*
   d. *"Sometimes I allow students to choose their own groups."*
   e. *"I feel it is necessary to group children myself. These formed groups remain the same for approximately 5-6 weeks, and then new ones are formed."*

2. What are some positive results you have had with "cooperative learning"?
   a. *"Even the students with the lowest achievements are able to feel they can contribute something."*
   b. *"Children learn to depend on each other, not the teacher. They accept individual as well as group responsibility. They all get involved."*
   c. *"They learn to use quiet voices, respect each other, discuss everything together, and come to one answer or conclusion."*
   d. *"They learn more from teaching others what they have learned."*
   e. *"The students seem to work longer and harder on projects. They are enthusiastic about participating in group activities, share responsibilities, and are proud to share the finished group project with the class."*

3. What are some negative results you have had with "cooperative learning"?
   a. *"Sometimes students just don't get along."*
   b. *"Some children dominate the situation, and others let the leaders do all the work."*
   c. *"Time management is a problem. Groups may be working well, but a 30–40 minute scheduled activity may take them 60–75 minutes to complete."*
   d. *"The noise level gets too high."*
   e. *"Kindergarten children may get a little unruly."*

1. Read the responses to survey question number 1. Which two responses have something in common? Which two responses represent opposing views?

   _____

2. Look at question number 2. Explain response *a* in your own words.

   _____

3. Look at response *c* to question number 2. What is another way of saying "quiet voices"?

   _____

4. Look at the responses to question number 3. Which responses are related to the personalities of the children in a group?

   _____

WRITING

Complete one of the following activities.

1. Write a story about a time when you were able to solve a problem with the help of a friend or relative. Describe how you felt at the time.
2. Write about an experience you have had studying with a friend. Include information that responds to the following questions: Was the work easier? Were you able to understand the task better? Did your grades improve?

Segment 9

Media Studies Would Help Kids Watch TV More Critically

From: *American Agenda*, 3/3/92
Begin: 1:20:14
Length: 4:14

BEFORE YOU WATCH

TALKING POINTS

Work in groups. Discuss your answers to the following questions.

1. In your opinion, do children in your home country spend more time watching television or reading books? How about in the United States?

2. Do you think that adults believe that what they see on television is real? Do children? Explain your answers.

3. Think of a sitcom (situation comedy) that you watch on television. What are some of the messages that the characters communicate to viewers regarding (a) sex, (b) violence, (c) housing, (d) clothing, and (d) the importance of having nice things?

PREDICTING

Based on the title of the video segment, *Media Studies Would Help Kids Watch TV More Critically*, what do you think you will see and hear on the video? Write down three items under each of the following headings. Then compare your answers with those of another student.

SIGHTS	WORDS
(things you expect to see)	(words you expect to hear)

1. _____   _____
2. _____   _____
3. _____   _____

KEY WORDS

The *italicized* words in the sentences below will help you understand the video. Study the sentences. Then write your own definition of each word.

1. Some schools have classes in reading and writing for *illiterate* adults.
   illiterate: _____

2. Some people *passively* accept every law that is passed, no matter how silly the law may be.
   passively: _____

3. Students who *actively* participate in their classes usually learn more.
   actively: _____

4. Air *surrounds* the earth.
   surround: _____

5. We see many *image*s on a television screen.
   image: _____

6. I need to lose ten pounds. I am *slightly* overweight.
   slightly: _____

7. Mammograms *detect* breast cancer at an early stage.
   *detect:*_____

8. My opinion regarding my children shows *bias* because I'm their mother!
   bias: _____

WHILE YOU WATCH

GETTING THE MAIN IDEA

Watch the news report and listen for the answers to the following questions. Take brief notes on the answers. Then compare your answers with those of another student.

1:20:20–
1:24:34

Who is doing **what, where?**
How are they doing this?

Who?	
What?	
Where?	
How?	

CHECKING YOUR PREDICTIONS

Look at the lists you made in the PREDICTING exercise on pages 81 and 82. Watch the video and check (✓) the items you actually see and hear.

1:20:20–
1:24:34

WHAT'S MISSING?

Listen to Peter Jennings' introduction to the news report. Fill in the missing words.

1:20:20–
1:20:35

Peter Jennings: On the *American Agenda* tonight, — how to watch

(1)_____ . We've put the (2)_____ on the

Agenda because by the time they leave school, American children

will have spent (3)_____ time in front of the

(4)_____ than in front of the (5)_____. So

they ought to know how it (6)_____. Our Agenda

reporter is Beth Nissen.

LISTENING FOR DETAILS

Watch the video. Circle the correct answers.

1:20:36–
1:21:58

1. Why are some American children illiterate?
   a. Because they don't know how to read.
   b. Because they don't go to school.
   c. Because they are watching television passively.
2. According to Professor Renee Hobbs of Babson College, in order to understand the unspoken messages on the screen,
   a. we must think critically.
   b. we must think uncritically.
   c. we must not think.

3. In which three countries are students being taught to think critically about what they see on television?
   a. Australia, Great Britain, and the United States.
   b. Great Britain, Australia, and Canada.
   c. Sweden, Norway, and Spain.
4. Which students in Ontario, Canada, are required to take media studies?
   a. Male students only.
   b. All high school students.
   c. All elementary students.
5. According to a student on the video, the women we see on television are
   a. thin.
   b. fat.
   c. slightly overweight.
6. According to Barry Duncan, the point of view most often presented on television indicates bias because
   a. it is the point of view of the general public.
   b. it is the female point of view.
   c. it is the male point of view.

TRUE OR FALSE?

Watch the video. Are the following statements *true* or *false?* Write **T** (true) or **F** (false). Make the false statements true by changing one or two words.

1:21:59–
1:22:38

1. _____ Use of a wide shot establishes a scene.

2. _____ Use of a wide shot also helps give someone authority.

3. _____ Use of a low angle shot helps give someone greater stature.

4. _____ A series of quick shots grabs the attention of the viewer.

5. _____ Editing is a powerful part of the media production process.

NOTETAKING

Watch the next part of the video and take brief notes on the answers to the following questions. Then compare your notes with those of another student.

1:22:39–
1:24:34

1. According to Mark, a student, how have media production classes changed the way he looks at television news, sitcoms, and commercials?

_____

_____

2. What are some of the things students in experimental classes at Oyster River Elementary School in Durham, New Hampshire, learn?

_____

_____

3. How can parents and teachers begin to train children to be more critical of what they see on television?

_____

_____

4. Explain what Beth Nissen means by saying that being critical "is a skill children will need to use all of their lives"?

_____

_____

MAKING TRUE SENTENCES

Watch the video. Then use the chart below to make six true sentences. Write the sentences below.

1:20:20–
1:24:34

Most American children . . .	1. are learning to be critical about what they see on television.
Students in other countries . . .	2. watch television passively.
	3. have few opportunities to take media literacy classes.
Many U.S. school districts . . .	4. can't afford to train teachers or buy equipment for media literacy classes.
	5. spend more time in front of the television than in front of the blackboard.

1. _____

2. _____

3. _____

4. _____

5. _____

LANGUAGE POINT: EXPRESSING CONTRADICTION

On the video, several ideas regarding what we should be teaching children about the media are presented and followed by a contradiction describing what is really happening. One way of *expressing a contradiction* is by using *but . . . not.* Study the following examples.

> We should watch TV critically, *but we do not.*

> All children should be given media classes, *but they are not.*

The ideas below are adapted from the video. Express contradictions by completing each sentence with the most appropriate ending from the box.

but we do not	but we are not
but it does not	but it is not
but they do not	but they are not

1. We should be disturbed by violent images, _____.

2. We should ask ourselves questions regarding opinions expressed,

   _____.

3. Children should be taught to think critically about issues raised on TV,

   _____.

4. Parents should watch TV with their children and ask them questions,

   _____.

5. The government should allocate more money to media literacy studies,

   _____.

6. Media literacy should be part of the curriculum in every American school, _____.

VOCABULARY CHECK

The following excerpts are from the video. What do the *italicized* idioms mean? Circle the correct answer.

1. And we uncritically *take in* those messages without thinking.
   a. put inside the house
   b. accept mentally
   c. write down
   d. play back on an answering machine

2. Students in other countries . . . are being taught in school to watch carefully and think critically about what *flashes past* on television.
   a. is briefly shown on the screen
   b. appears when a light goes on
   c. has happened in the past
   d. disappears when a light goes on

3. Here's a guy who's *larger than life.*
   a. has lived for a long time
   b. is a great hero
   b. is very kind to people
   d. is very tall

4. I always try and figure out . . . *where they're coming from.*
   a. why they're leaving
   b. what their nationality is
   c. what their attitude is
   d. where they live

5. I always try and figure out . . . *what their angle is.*
   a. what message they want people to get
   b. what they look like in front of the camera
   c. how much money they want people to pay
   d. what camera shots they use

WORD PUZZLE

Unscramble the letters to make a word to fit the definition. All the words are used in the video.

1. A prejudice or a strong opinion (saib): ☐○☐○
2. A process of preparing a film, video, etc., by deciding which parts to cut out or leave in (tidieng): ☐☐☐○☐○☐
3. A single picture on a role of film (meraf): ☐☐☐☐○
4. A person who is watching TV (wireev): ○☐☐☐☐☐
5. A writing slate used in classrooms (aldkbcbroa): ☐○☐☐☐☐○☐☐☐
6. A test or research used for studying new methods (mexnepriet): ○☐☐☐☐○☐☐☐☐

Rearrange the letters in the circles to form a word that the word *this* refers to in the sentence from the video that appears below.

> *American children spend more time in front of **this** than in front of the blackboard.*
> The __ __ __ __ __ __ __ __ __ __

DISCUSSION

Work in groups. Discuss your answers to the following questions.

1. In your opinion, how much influence does television have on your life—very little? some? a lot? Explain your answer.

2. Give an example of what you consider to be a good TV program. How often do you get to watch programs like this?

3. What kinds of things should parents and teachers do to ensure that children will not be badly influenced by television?

4. Should children be given a limited amount of time to watch TV? If so, how much time should they be given? What time of day is most appropriate for them to watch? Discuss both weekends and weekdays.

5. Do you think media classes are worthwhile? If so, how would you introduce this concept in your local school system?

ROLE PLAY

Work in pairs. One student will play the role of the interviewer. The other will play the role of the professor of media studies. Read the situation and the role descriptions below and decide who will play each role. After a ten-minute preparation, begin the interview.

THE SITUATION: **A TV Interview**

A TV program called "Educational Innovations" has invited a professor of media studies to talk about the importance of media education and how children can be taught to think critically about what they see on television.

ROLE DESCRIPTION: **Interviewer**

You are the TV interviewer. Prepare a list of questions to ask the professor about the importance of media education and what parents and teachers can do to teach children to think critically about what they see on television.

ROLE DESCRIPTION: **Professor of Media Studies**

You are the professor. Be prepared to answer questions about the importance of media education and what parents and teachers can do to teach children to think critically about what they see on television.

READING

Read the following article and then answer the questions that follow.

MEN HAVE A RIGHT TO FAIR TREATMENT, TOO

Much has been said and done about giving fair treatment to women on television. In early TV shows, women were usually shown as housewives who spent their time cooking and cleaning, thus demeaning the challenging role of motherhood. When women were shown in the workplace, they were most often depicted in positions of lower stature than males. In TV commercials, a woman's opinion was often presented as frivolous and uneducated, giving the viewer the impression that women were less intelligent than men.

These images changed dramatically in the '80s. Women were still shown as homemakers, but were more often shown sharing housework and parental responsibilities with their male partners. In the workplace, they appeared in important managerial positions, just like their male counterparts. Women's decisions were respected as being intelligent, thereby erasing the previous negative image. But while the female image in the media has taken a turn for the positive, what have the media done to ensure that the male image is equally fair?

Let's examine a commercial.

In the opening scene we see a split screen. On one side, a young man is opening the door to a kitchen cabinet. On the other, a young woman is doing the same. Each is faced with the decision of choosing either Brand A or Brand B cereal for breakfast.

The young woman chooses Brand A (the brand the advertiser is trying to sell), which contains all the essential minerals and vitamins she needs for stamina during a work day. The young man chooses Brand B (the competitive cereal), which of course provides him with no essential ingredients. Later that morning, both are seen giving a competitive sales presentation to an executive board. The female sales agent triumphantly gets the contract because she was smart enough to choose the right cereal for breakfast.

The negative image of men that is transmitted by the media is not limited to commercials showing males making unintelligent choices. In the situation show *Roseanne*, the star of the show, Roseanne Barr, has made a career of saying things like "You may marry the man of your dreams, ladies, but 15 years later you're married to a reclining chair that burps." In one episode of the show, she says to her TV show son, "You're not stupid. You're just clumsy like your daddy."

In dramatic programs, sleazy criminals, and incompetent characters are often depicted as males. In a TV series of cops vs. the bad guys, young women's roles are upgraded to those of assertive, demanding police officers. But the low-down characters — drug dealers and thieves — are usually played by young men.

As an adult viewer, one wonders if it is really necessary for men to launch the same defensive tactics that women used in the '80s to ensure equal treatment in the media. Wouldn't it be easier and fairer if the media simply treated each sex with respect? If the television doesn't accept this responsibility, what hidden messages about male roles are we passing on to the next generation? Don't men have a right to fair treatment too?

1. How were women portrayed in early TV shows and commercials?

   _____

2. How did this image change in the '80s?

   _____

3. What did the man do wrong in the commercial that is described?

   _____

4. What roles are played by young men in TV series of cops vs. the bad guys?

   _____

5. According to the author, what responsibility should the media accept?

   _____

6. What hidden messages regarding male roles are we passing on to the next generation?

   _____

7. According to the article, are men and women treated equally on TV?

   _____

8. What is the main idea of this article?

   _____

WRITING

Complete one of the following activities.

1. You are a parent of a child who is in elementary school. You think your child's school should include media classes in its curriculum. Write a letter to the principal of your child's school and explain why.
2. Write a 150–200 word description of your favorite TV program. Include the name of the program and the type of program it is. Explain why you enjoy the program so much.

Segment 10
A Dinosaur Named Sue

From: *Prime Time Live*, 1/7/93
Begin: 1:24:35
Length: 11:39

BEFORE YOU WATCH

TALKING POINTS

Work in groups. Discuss your answers to the following questions.

1. Scientists often study animals to learn about the past. What have they learned about the conditions on Earth in the time of the dinosaurs?

2. As a child, you might have read some stories about how dinosaurs became extinct. Explain any theories you remember. Mention any new information you might have learned recently.

PREDICTING

The video is about the discovery of a skeleton of *Tyrannosaurus rex*, a type of dinosaur. Write your answers to the following questions. Then compare your answers with those of another student.

1. What do you know about *Tyrannosaurus rex*?

_____

_____

2. What are you unsure of about *Tyrannosaurus rex*?

_____

_____

3. What do you expect to learn from the video?

_____

_____

KEY WORDS

The words and phrases below are used on the video. Which will be used to name body parts of dinosaurs? Which will be used to describe dinosaurs? Put each word or phrase into one of the categories on the chart. Then compare your chart with that of another student.

cheekbone	lower jaw	shoulder blade
creature of mystery	meat-eater	skull
facial bones	prehistoric	three-stories high
fearsome	ruled the earth	vertebrae

**Words Used to Name
Body Parts of Dinosaurs**

**Words Used to
Describe Dinosaurs**

_____

_____

_____

_____

_____

_____

WHILE YOU WATCH

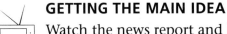

GETTING THE MAIN IDEA

Watch the news report and listen for the answers to the following questions. Take brief notes on the answers. Then compare your answers with those of another student.

1:24:43–
1:36:22

> **Who** is Sue?
>
> **What** is the *battle* about?
>
> **When** and **Where** did the *battle* begin?
>
> **Why** has it become the *custody battle of the century*?
>
> **Why** are the findings so important?

Who?	
What?	
When?	
Where?	
Why?	
Why . . . Important?	

CHECKING YOUR PREDICTIONS

Look at your answer to question 3 in the PREDICTING exercise on page 91. What did you learn from the video? What did you expect to learn?

1:24:43–
1:36:22

WHAT'S MISSING?

Listen again to Sam Donaldson's introduction to the news story. Fill in the missing words.

1:24:47–
1:25:13

Sam Donaldson: It may be the (1) _____ battle of the century.

(2) _____ , Indians, the U.S. government — they're all

trying to lay (3) _____ to a bunch of old bones. Well,

not just any old bones. These bones happen to be (4)_____million

years old. In the midst of it all is the man who (5) _____

them up in (6)_____. For him, it was the

find of a (7) _____. Now he could wind up in

(8)_____.

Sylvia Chase has the amazing tale of the (9)_____

over a dinosaur named (10) _____.

LISTENING FOR DETAILS

Watch the next part of the video. Circle the correct answers.

1:25:15–
1:30:31

1. Why may Sue be the greatest find in the history of dinosaur hunting?
   a. She's the only example of a *Tyrannosaurus rex*.
   b. She's the most complete example of a *Tyrannosaurus rex*.
   c. She's the oldest example of a *Tyrannosaurus rex*.

2. What did Sue probably die from?
   a. A bite from another dinosaur.
   b. Starvation.
   c. Both a and b.
3. What is the most important thing about Sue?
   a. That she was a vegetarian.
   b. That she lived to be 100 years old.
   c. That 90 percent of her skeleton was recovered.
4. What parts of Sue are complete or nearly complete?
   a. Her tail and her shoulder blade.
   b. Her arm and her skull.
   c. Both a and b.
5. To which institution did Pete Larson say he was going to donate Sue?
   a. Harvard University.
   b. A non-profit museum in Hill City.
   c. The Smithsonian.
6. Which of the following is NOT true about Hill City?
   a. It's a tourist town.
   b. Most of its population is very wealthy.
   c. The town hoped the dinosaur museum would attract more tourists.
7. What happened in Hill City on the morning of May 14?
   a. Federal agents arrived to take away Sue's bones.
   b. One of the townspeople had a nightmare about Sue.
   c. The dinosaur museum was open to the public for the first time.
8. How much is Sue worth?
   a. $20,000.
   b. $200,000.
   c. $20,000,000.

NOTETAKING

Watch the next part of the video. Reporter Sylvia Chase asks three people the same question, "Who owns Sue"?, and gets three different answers. What answer does each person give?

1:30:15–
1:30:50

1. Maurice Williams: _____

   _____

2. Kevin Schieffer:_____

   _____

3. Pete Larson: _____

   _____

WHO'S WHO?

Look at the chart below. Then watch the video and check (✓) the appropriate boxes.

1:29:15–
1:32:15

Who . . .?	Gregg Bourland	Drue Vitter	Maurice Williams	Kevin Schieffer
1. is the mayor of HIll City, South Dakota				
2. is the owner of the ranch where Sue was found				
3. is the chairman of the Cheyenne River Sioux tribe				
4. is an attorney for the U.S. government				

PUTTING EVENTS IN ORDER

Read the sentences below. Then watch the video and put the events in order in which they occurred. Number them 1 to 6. The first event has been numbered for you.

1:30:51–
1:33:59

____ Larson dug up Sue's bones.

_1_ Pete Larson paid Maurice Williams $5,000 for permission to excavate the fossil from land on Williams' farm.

____ The Cheyenne River Sioux tribe proposed a partnership with Pete Larson.

____ The bones were locked up in a steel box at the South Dakota School of Mines.

____ The U.S. government sent FBI agents to seize the bones.

____ Protesters demanded Sue be restored to Larson.

TRUE OR FALSE?

Watch the last part of the news story. Are the following statements *true* or *false*? Write **T** (true) or **F** (false). Make the false sentences true by changing one or two words.

1:34:00–
1:36:22

1. ____ The government spent $100,000 to restore Sue.

2. ____ Kevin Schieffer says Larson went on federal land and stole federal property.

3. _____ Larson invited scientists to examine Sue.

4. _____ It is believed that Larson will be indicted on major felony charges.

5. _____ Larson has won money in his fight with the government.

6. _____ Schieffer is willing to tell the public exactly how much money he has spent on the case so far.

7. _____ Sue's bones are still locked up in a steel box.

8. _____ Larson believes that Sue is being destroyed.

AFTER YOU WATCH

LANGUAGE POINT: PASSIVE PAST TENSE

On the video, Sylvia Chase uses the *passive past tense* when she says, "Sue was taken from the privately held Williams ranch." In general, we use the *passive past* when we want to emphasize the result of an action, rather than the performer of the action. *To form the passive past tense, use* **was** *or* **were** *+ past participle.* Complete the paragraphs that follow with the passive past form of the verbs in parentheses. The first one has been done for you.

Several years ago, some old dinosaur bones ____*were found*____ on a farm
(1. find)

in Hill City, South Dakota. The excavation to remove the bones

_____ by Pete Larson, a dinosaur hunter whose company
(2. conduct)

specializes in the delicate task of separating fossils from rock. Altogether,

about 90 percent of the dinosaur's skeleton _____ , and the
(3. recover)

bones_____ to be about 65 million years old. The dinosaur
(4. judge)

_____the name "Sue," after Sue Hendrickson, the paleontologist
(5. give)

who led Larson and his crew to the bones.

Shortly after Larson announced that he planned to give Sue to a non-

profit museum in Hill City, Hill City _____ by FBI agents
(6. invade)

who claimed that Sue was a public treasure and the property of the

U.S.government. When it _____that Sue was worth about
(7. announce)

$20 million, a lot of other people, including the owners of the ranch from

which Sue's bones _____ , began to claim that Sue belonged to
(8. remove)

them. Finally, Sue _____ to the South Dakota School of Mines
(9. move)

where she _____ up in a steel box.
(10. lock)

VOCABULARY CHECK

The *italicized* words are used on the video. Cross out the word or phrase in
each row that *does not* have a similar meaning to the word in *italics*.

1. *remote*	nearby	distant	far off
2. *insight*	knowledge	awareness	misunderstanding
3. *reputation*	fame	guilt	status
4. *seize*	take	capture	release
5. *looting*	donating	stealing	robbing
6. *fray*	conflict	agreement	battle
7. *reprehensible*	innocent	at fault	guilty
8. *authentic*	fake	real	genuine
9. *indicted*	charged	set free	arrested
10. *decompose*	decay	improve	rot

DISCUSSION

Work in small groups. Discuss your answer to the following questions.

1. Many of us may never have found a fossil. Nevertheless, we might
   have something in our families that is very old, is considered a prized
   possession, and is not for sale. Do you have such an item in your
   family? If so, describe the item and its background.

2. On the video we saw government agents seizing something they
   claimed as theirs although many people felt they were wrong. Has
   the government ever taken something personal from you? If so,
   describe the incident and how you felt.

3. At the end of the video, we are given the sense that the battle over Sue
   is still going on in the courts and that it may be a long time before the
   issue is resolved. Have you ever had to deal with the court system in
   the United States or in another country? If so, describe the case and
   how long it took for it to be resolved.

ROLE PLAY

Divide the class into four groups. One group will play the role of the environmentalists, another will play the role of the concerned citizens, the third will play the role of the government officials, and the fourth will play the role of the museum's board of trustees. Read the situation and the role descriptions and decide which group will play each role. After a ten-minute preparation, begin the role play.

THE SITUATION: **A Court Battle in Hill City**

There is an on-going court battle in Hill City to determine who is the rightful owner of Sue. Various groups have been called to give their opinions about why they feel their organization should be entitled to profits from the revenues earned by the museum.

ROLE DESCRIPTION: **Environmentalists**

You are worried about the environmental damage that will be caused by large groups of tourists flocking to the area in cars and trucks. You want the profits to be used to protect plants and animals in the area from the added pollution that this tourism would cause.

ROLE DESCRIPTION: **Concerned Citizens Against Crime**

You are worried about the increased crime that would result from an increase in population, tourists, and businesses. You feel some money should be set aside to pay for more police officers, another police car, and the organization of a volunteer crime prevention force.

ROLE DESCRIPTION: **Local Government Officials**

You would like to beautify Hill City. You would use the profits to add a new park, put up benches around town, and renovate the old Court House, which is a symbol of the town. You also think some of the money should be used to put up a sign reading "Welcome to Hill City — Sue's Hometown" at the entrance to the town.

ROLE DESCRIPTION: **The Museum's Board of Trustees**

Your group helped to finance Pete Larson's excavation. You want a share of the profits in order to modernize the museum and build a new wing dedicated only to Sue and the history of the dinosaurs.

READING

Read the following article to learn about some innovative ideas concerning dinosaurs. Then answer the questions that follow.

A NEW VIEW OF DINOSAURS

After dominating life on Earth for about 140 million years, dinosaurs disappeared 65 million years ago. Until as recently as the 1970s, most people, including the experts, believed that dinosaurs were solitary, dim-witted, and slow-moving beasts. However, fossil discoveries and interpretations in the last quarter of the 20th century have led scientists to take a new view of these massive creatures.

Certain similarities between dinosaurs and birds, who are believed to be the descendants of the dinosaurs, have helped to create a new view of the prehistoric creatures. Studies indicate that certain species of dinosaurs, such as the Allosaurs, ran in quick, birdlike steps. Dinosaur expert Robert T. Bakker of the University of Colorado has even claimed that dinosaurs, like birds and mammals, were warm-blooded animals.

There is also evidence to suggest that dinosaurs cared for their young in the same way that birds do. John Horner of Montana State University found the nests of two different dinosaur species in northern Montana. The teeth of baby dinosaurs still in the nests showed signs of wear, which suggests that adult dinosaurs brought them food, just as adult birds brings food to their young today.

Robert T. Bakker also believes that dinosaurs possessed a social structure by which the older, larger creatures protected the younger, smaller ones. In the 1930s Roland T. Bird found footprints of twenty-four Sauropods (four-legged, plant-eating dinosaurs) on a ranch in Texas. More recently, Bakker has argued that these footprints are evidence that the larger animals ran along the outside of the herd, protecting the smaller, younger ones who ran inside the group.

Why did the dinosaurs disappear? Their extinction has been attributed to many causes. Some experts believe they died out when the earth darkened by the impact of a catastrophic asteroid that fell from outer space. Others claim that gradual changes in climate were responsible. Still others put the blame on increased volcanic activity. We may never know exactly what caused dinosaurs to become extinct, but scholars continue to paint a clearer and fuller picture of what these amazing creature were like.

1. For how long have dinosaurs been extinct?

   _____

2. What did people use to believe dinosaurs were like?

   _____

3. When did this older view of dinosaurs begin to change? What caused the change?

_____

4. What three possible similarities between dinosaurs and birds are mentioned in the article?

_____

5. What does Robert T. Bakker claim is proof that dinosaurs possessed a social structure?

_____

6. What three things are mentioned as possible causes of the extinction of dinosaurs?

_____

WRITING

Complete one of the following activities.

1. You live in Hill City, South Dakota. Write a brief letter (150–200 words) to the editor of the local newspaper, *The Hill City Times*, giving your views of the battle over Sue's bones and what you think should be done with her skeleton.

2. Imagine you are back in the age of the dinosaurs. The earth is abundant with plant life. Describe what you see, how it smells, and what you feel. Describe the dinosaurs that pass by. What are they doing? Where do you think they are going?

Segment 11
Joey's Best Friend

From: *Prime Time Live*, 1/30/92
Begin: 1:36:25
Length: 10:12

BEFORE YOU WATCH

TALKING POINTS

1. What is physical therapy? Give a few examples in which physical therapy would be required.

2. Do you think that animals could be involved in physical therapy? If so, explain how you think this might work.

3. Do you think that it is possible for animals to communicate emotion? If so, describe personal situations in which you have had evidence of this.

PREDICTING

Work in groups. The news report is about a research center where a select group of properly trained human visitors are allowed to swim with dolphins. What do you think you will see and hear on the video? Write down five items under each of the headings below. Then compare your answers with those of another student.

<table>
<tr><td align="center">SIGHTS
(things you expect to see)</td><td align="center">WORDS
(words you expect to hear)</td></tr>
<tr><td>1. _____</td><td>_____</td></tr>
<tr><td>2. _____</td><td>_____</td></tr>
</table>

3. _____   _____

4. _____   _____

5. _____   _____

KEY WORDS

The *italicized* words below will help you understand the video. Study the definitions. Then use each word in a sentence of your own.

1. *dolphins*: fish-like mammals, gray in color and with a beak-like snout, that live in the sea

   _____

2. *impaired*: unhealthy or defective because of illness or disease

   _____

3. *congenital*: (of diseases, etc.) present from or before birth

   _____

4. *stroke*: a sudden illness which affects part of the brain and which can cause loss of ability to move, speak clearly, etc.

   _____

5. *limp*: having no strength or energy

   _____

6. *ordeal*: extremely unpleasant and difficult experience

   _____

7. *petrified*: extremely frightened or fearful

   _____

8. *handicapped*: suffering from a physical or mental disability

   _____

WHILE YOU WATCH

GETTING THE MAIN IDEA

Watch the news report and listen for the answers to the following questions. Take brief notes on the answers. Then compare your answers with those of another student.

1:36:32–
1:46:54

Who is Fonzie?

What have scientists recently discovered about dolphins?

Where is the research being done?

Why is this discovery so important?

Who?	
What?	
Where?	
Why?	

CHECKING YOUR PREDICTIONS

Look at the lists you made in the PREDICTING exercise on page 101 and 102. Watch the news report again and check (✓) the items you actually see and hear on the video.

1:36:32–
1:46:54

WHAT'S MISSING?

Listen to John Quinones' introduction to the news report. Fill in the missing words.

1:37:09–
1:38:00

John Quinones: For more than 50 million years, long before man ever

walked the earth, they have graced this dark (1)_____

with majestic beauty. For decades we've known them as the most

(2) _____ creatures of the sea, next on the intellect scale

to humans and primates. It now turns out they may be smarter

and perhaps more (3)_____ than we thought.

Scientists at Dolphins Plus, a research center in the Florida Keys,

have (4)_____ that (5)_____ display a

surprising array of humanlike (6)_____, like joy and

(7)_____, particularly for children who are physically

or mentally (8)_____, children like five-year-old Joey

Hoagland.

TRUE OR FALSE?

1:38:00–
1:40:18

Watch the video. Are the following statements *true* or *false*? Write **T** (true) or **F** (false). Make the false sentences true by changing one or two words.

1. _____ Joey acquired truncus arteriosis after he was born.

2. _____ The left side of Joey's body was paralyzed after a stroke.

3. _____ While he was in a coma in the hospital, Joey could speak and eat.

4. _____ Doctors felt confident that Joey would pull out of the coma.

5. _____ Joey did not cooperate with doctors after the coma because he was afraid.

6. _____ Dolphin Plus is a research center that uses dolphins as part of its physical therapy program.

7. _____ The dolphins at Dolphins Plus live in captivity.

8. _____ Betsy Smith is a sociologist who has a retarded brother.

LISTENING FOR DETAILS

1:39:52–
1:45:03

Watch the video. Circle the correct answers.

1. How long ago did Betsy Smith notice that swimming with dolphins increased mental agility in the retarded?
   a. 10 years ago.
   b. 13 years ago.
   c. 15 years ago.

2. Why does Ms. Smith feel that dolphins display "altruistic behavior"?
   a. Because they care for their elderly and their sick.
   b. Because they are strong swimmers.
   c. Because they are 700 pound mammals.

3. How did Joey act when he first saw the dolphins?
   a. He acted strong and confident.
   b. He seemed weak and uninterested.
   c. He was shy and apprehensive.

4. According to Chris Blakenship, what did Fonzie provide that medicine couldn't?
   a. Physical therapy.
   b. Emotional therapy.
   c. Both a and b.

5. What have some studies shown about dolphins and stress?
   a. Dolphins display a powerful stress-reducing effect on children.
   b. Dolphins are so powerful that they increase stress levels in adults.
   c. Dolphins neither increase or decrease stress level in people.

6. What does the study completed by Dr. John Schull and David Smith reveal about dolphins?
   a. Dolphins can distinguish between a high- and a low- pitched tone.
   b. Dolphins cannot distinguish between a high- and a low- pitched tone.
   c. Dolphins make mistakes and therefore cannot really distinguish tones.
7. What does the study conclude?
   a. A dolphin's brain is too small to think.
   b. A dolphin can think and feel.
   c. A dolphin can feel, but it cannot think.
8. What allows dolphins to be so aware of human needs?
   a. They have remarkable sonar capacity.
   b. They can think and figure out simple mathematical problems.
   c. They can sing.

CHECKING WHAT YOU HEAR

On the video, John Quinones says, "Dolphins have been known as the most intelligent creatures of the sea. It now turns out that they may be smart and perhaps more sensitive than we thought." Listen to the video for reasons why dolphins work so well with handicapped children and the elderly. Check (✓) the appropriate boxes.

1:40:19–
1:44:14

Who says that *dolphins*. . .?	Chris Blackenship	Betsy Smith	Dr. John Schull	John Quioness
1. relax and become very different —very, very, quiet and calm, and will work with a handicapped child in the water for as long as it takes				
2. care for the sick, ailing, and elderly				
3. have an aspect that no one yet really understands				
4. sense when a person is ill				
5. display a powerful stress-reducing effect on children				
6. are thinking, feeling, perhaps self-aware animals, and you realize that they have a much deeper kind of relationship with you				
7. may be able to use their sonar to perceive things that are going on within the skin				

LANGUAGE POINT: POSSESSIVES

One of the easiest ways to show possession or that something belongs to someone is by *adding an apostrophe (') to a noun or a name* as in the title of this segment, *Joey's Best Friend* . As we watch the video we realize that Fonzie, a 700-pound dolphin, is the friend that belongs to Joey. We can also show possession by *using a possessive adjective before a noun*. For example, another way of saying "Joey's Best Friend" is "His Best Friend." Here the possessive adjective "His" takes the place of the word "Joey's."

Rewrite each of the sentences below, substituting one of the possessive adjectives in the box for each noun or name that has an apostrophe. The first one has been done for you.

POSSESSIVE ADJECTIVES

my   your   his   her   its   our   their

---

1. Joey's best friend is Fonzie.

   His best friend is Fonzie.

2. The doctors' opinion was that Joey would not make it.

   _____

3. Joey's parents were afraid that Joey wouldn't trust them.

   _____

4. Betsy Smith's observation revealed that dolphins work very well with children.

   _____

5. Whatever interaction there is with humans is strictly at the dolphins' discretion.

   _____

5. The boy's motor skills have increased tremendously.

   _____

VOCABULARY CHECK

The *italicized* words are used on the video. Cross out the word or phrase that *does not* have a similar meaning to the word in *italics*.

1. *behavior*	place	conduct	actions
2. *ailing*	ill	healthy	sick
3. *elderly*	aged	old	young
4. *apprehensive*	afraid	anxious	fearless
5. *nuzzled*	made a noise	snuggled	rubbed with the snout
6. *toss*	catch	throw	pitch
7. *paralyzed*	flexible	numbed	unable to move
8. *feed*	nourish	take away	give food to

DISCUSSION

Work in groups. Discuss your answers to the following questions.

1. Joey weighs 40 pounds. Fonzie weighs 700 pounds. If Joey had been your child, a close friend, or relative, would you have allowed him — a five-year-old child — to swim with a dolphin?

2. Joey was in a coma for a while. When he woke up, he was petrified and didn't trust humans. Do you know anyone who has been in a coma and has come out of it? Tell about the experience.

3. The dolphins at Dolphin Plus do not need to perform tricks for a crowd or earn their food by performing. However, many other animals in captivity are expected to do that. How do you feel about this issue?

ROLE PLAY

Work in pairs. One student will play the role of the director of a new scientific laboratory center that will use animals for laboratory research. The other student will play the role of the pro-animal activist. Read the situation and the role descriptions below and decide who will play each role. After a ten-minute preparation, begin the interview.

THE SITUATION: **A Confrontation on Camera**

A new scientific laboratory has been funded by the government to start operating in the state of Florida. The lab will study the behavior of dolphins to find out how this interesting species of mammal can be used to better serve humans. A well-known animal activist and some people of the area are protesting the existence of such a lab in their area. They have called in a TV crew to record the confrontation between the animal activist and the lab director, in order to make the public aware of their displeasure.

ROLE DESCRIPTION: **Lab Director**

You grew up in Florida near the ocean and have always been fascinated by dolphins and their behavior. As an oceanographer, you have learned many facts about the species that have sparked your interest even further. You would never do anything to hurt the dolphins that will be under your care. You will simply head a team of scientists in their research on how dolphins could better help humans. The dolphins will be tagged and enclosed in a pool area until they become accustomed to people. Once a relationship is established between the researchers and the dolphins, the dolphins will be allowed to roam in the nearby ocean, the same way that people let out pets they know will return.

ROLE DESCRIPTION: **Animal Activist**

For a long time you have been fighting the efforts of government and private industry to constrain wild animals. You think any animal that was born free should remain free. You feel that tampering with nature will prevent animals from using their natural instincts to hunt for food, take care of themselves, and reproduce. You don't believe that humans have the right to make other species subservient to them. Dolphins don't put humans in natural habitats and study the ways that humans can serve them, so what gives humans the right to do that to another species?

READING

Read the following article to find out about a mother and son's experience with a dolphin. Then answer the questions that follow.

"MOM, THERE'S MR. DOLPHIN!"

It was a bright, cool morning, the last day of our four-day mini-vacation on the southwest coast of Florida. Armed with a shovel, pail, towel, suntan lotion, and all the other amenities required to keep a five-year-old entertained at the seashore, I headed for the beach with my son Ector trailing not far behind. Erik, his older brother, had opted to stay by the pool area, where he was entertaining himself with some newly made friends. As I sat on the beach chair and stared out into the Gulf, for perhaps the last time during this vacation, I experienced a feeling of sadness. I had always felt a connection with the sea, which had a tremendous effect on me. Just staring at it helped me remember to keep things in perspective The realization that this morning represented a kind of farewell, at least for the moment, seemed to require something more formal than simply witnessing Ector build another sandcastle on the beach. I needed to be closer to the sea. I wanted to become part of it and take that feeling back with me and cherish it until the next time that we would meet again.

Usually, hotels by the shore have sailboats and catamarans for rent. Although I had had very little experience navigating, I felt confident enough to take charge of a catamaran for the half hour it would belong to us. The cabin boy gave me some last-minute instructions and reassured me that there would be no danger as long as I stayed closed to shore. He helped us put on our life-jackets and before long we were sailing on the bay.

The waters were calm, and the sea breeze soon carried us out of the immediate beach area. Ector was delighted and splashed the water with his left hand as he lay flat on this stomach and sang familiar nursery tunes. I was determined to stay close to shore and was maneuvering a northeast turn when a cross-wind offered some resistance. Trying not to alarm my son, I decided to surrender to the wind and make a southeast turn instead. During this maneuver I heard Ector say, "Mom, there's Mr. Dolphin!"Concerned with successfully completing the turn, I was only able to catch a glimpse of a small massive gray shadow in the water just ahead of us. Ector began calling out to him the same way one would call a pet. I was pleased to see that my son had such an affinity for animals, and I did not want to take away his hope that the dolphin would answer his pleas. To my amazement the gray shadow reappeared, this time on the left-hand side of the ship where Ector was lying. Concerned for Ector's safety, I asked him to move to the center. He obeyed, but continued to talk to Mr. Dolphin as if the dolphin understood. This time Mr. Dolphin swam straight through the center of our catamaran. As he made a swift turn ahead of us, he leapt from the water high enough to allow us to witness the full size of his body. I was amazed to realize that this beautiful creature seemed to actually be interacting with my son. Ector excitedly cried out to the dolphin, splashing his little hand in the water as he pleaded with the dolphin to let him touch it. Once again the dolphin swam under the boat, and for a brief moment there was contact between the innocence of child and the magnificence of a creature from the sea.

Our half-hour ride was almost up. I had to turn toward the shore. Ector waved good-bye to his new-found friend. He hugged me tightly and said, "He let me touch him, mummy. He really did!" In a very subtle way, I felt Ector too had been touched by the mysteries of the sea and its creatures. Perhaps he also would feel the same connection I feel with the sea, and now we both had an experience to share and cherish.

1. How many days had the family been at the seashore?

   _____

2. How did the mother feel about leaving?

   _____

3. Why did she decide to rent a catamaran?

   _____

4. Did she have any problems steering the boat? Explain your answer.

   _____

5. What did the mother ask the little boy to do to guard his safety?

   _____

6. What did the dolphin do that led the mother to believe that there was
   an interaction going on between her son and the dolphin?

   _____

7. What was the most important moment of the interaction?

   _____

8. In your opinion, did this experience make the mother and the child
   feel closer? Explain your answer.

   _____

WRITING

Complete one of the following activities.

1. As a pet owner, you know that animals can sense and feel. You believe
   that such positive communication could be beneficial to sick people,
   who need to both give and receive love and care. Write a letter to the
   Pro-Care Animal Corporation, located at 2477 North Warrington Street,
   New York, New York 10023, explaining the need for interaction
   between animals and humans.
2. Do you believe that animals should be used for scientific experiments?
   Write a letter to the editor of an English-language newspaper, giving
   your opinion as to whether animals should be used by scientists in
   experiments. Include at least three reasons why you feel the way that
   you do.

Segment 12
Environmentally Friendly Design

From: *American Agenda,* 8/26/93
Begin: 1:46:57
Length: 4:29

BEFORE YOU WATCH

TALKING POINTS

Work in groups. Discuss your answers to the following questions.

1. What source of power (gas? electricity? other?) do you use to keep your home warm or cool? Is it efficient? Is it costly? How could you change it to make it more efficient and less costly?
2. Have you ever been in a house or other building with an "environmentally friendly" design? What materials were used to build it? What energy-saving features did it have? How do these save money as well as energy?
3. Most of us can do a little every day to help save water, conserve energy, and assure ourselves of more fresh air. What, if anything, do you do to help the environment?

PREDICTING

Work in groups. Based on the title of the news report, write down three questions you think will be answered on the video.

1. _____

2. _____

3. _____

KEY WORDS

The *italicized* words in the sentences below will help you understand the video. Study the sentences. Then match the words with the definitions.

1. The *architects* showed us the plan for the new office building.
2. Environmentalists emphasize the *interdependence* of all living things.
3. The heating systems in some old buildings *waste* a lot of energy.
4. Gas fumes from cars and trucks are the cause of the *polluted* air in most major cities.
5. The company *recycled* glass, plastic, and paper to help save the environment.
6. Some buildings have electronic *sensors* that turn lights on and off.
7. New *technologies* make it easier and cheaper for people to communicate over long distances.

1. _____ *architects*   a. use more than is necessary
2. _____ *interdependence*   b. treated in order to use again
3. _____ *waste*   c. instruments that react to certain physical conditions
4. _____ *polluted*   d. people who design buildings
5. _____ *recycled*   e. scientific methods and equipment
6. _____ *sensors*   f. dirty and dangerous to use or live in
7. _____ *technologies*   g. the condition of people or things all depending on each other

WHILE YOU WATCH

GETTING THE MAIN IDEA

1:47:04–
1:51:33

Watch the news report and listen for the answers to the following questions. Take brief notes on the answers. Then compare your answers with those of another student.

Who is doing **what**? **Where**?

Why and **how** are they doing this?

Who?	
What?	
Where?	
Why?	
How?	

CHECKING YOUR PREDICTIONS

Look at the questions you wrote in the PREDICTING exercise on page 111. Watch the video. Which of your questions are answered in the video? What answers are given?

1:47:04–
1:51:33

WHAT'S MISSING

Listen to Diane Sawyer's introduction to the news report. Fill in the missing words.

1:47:04–
1:47:25

Diane Sawyer: We've put the (1) _____ of the future on the

American Agenda tonight. In Chicago this summer, 5,000

(2)_____ from all over the country (3) _____

what they called a (4) _____ of (5) _____

a pledge to (6) _____ buildings that are kinder to the

(7)_____. As our Agenda reporter Barry Serafin tells

us, some buildings like that already (8) _____.

TRUE OR FALSE?

Watch the video. Are the following statements *true* or *false?* Write **T** (true) or **F** (false). Make the false sentences true by changing one or two words.

1:47:25–
1:49:10

1. _____ A lot of city buildings waste energy.

2. _____ There is a cultural shift toward designing buildings that help to solve environmental problems.

3. _____ The new headquarters of the National Audubon Society in New York is a recycled old building.

4. _____ Architects wasted hundreds of tons of steel and concrete on the new headquarters.

5. _____ Randolph Croxton is one of the architects who designed the new headquarters.

6. _____ Sensors in the headquarters can detect when an office is empty or occupied.

7. _____ The new lighting system cost an additional $100,000.

8. _____ The new lighting system is less efficient than in most offices.

9. _____ The carpeting in the building is made of polyester.

10. _____ The air in the building is cleaned and recirculated six times a day.

NOTETAKING

Watch the video and take brief notes on the answers to the following questions. Then compare your notes with those of another student.

1:49:11–1:49:42

1. What is Jan Beyea's opinion of the National Audubon Society's new headquarters?

   _____

2. How many new products and technologies has a company called Environmental Construction Outfitters assembled?

   _____

3. What product does Environmental Construction Outfitters make from recycled plastic jugs?

   _____

4. What material does the company use to make insulation ?

   _____

CHECKING WHAT YOU SEE

Watch the next part of the video with the *sound off*. Check (✓) the things that you see.

1:49:51–1:51:02

1. ❑ Wal-Mart store
2. ❑ an environmental education center
3. ❑ a mini-zoo
4. ❑ some skylights
5. ❑ a playground
6. ❑ some benches
7. ❑ some bins
8. ❑ a hose
9. ❑ parking lot
10. ❑ a package recycling sign

LISTENING FOR DETAILS

Now watch the video with the *sound on*. Circle the correct answers.

1:49:51–1:50:47

1. In which middle-American city is the new Wal-Mart store located?
   a. Lawrence, Massachusetts.
   b. Lawrence, Arkansas.
   c. Lawrence, Kansas.

2. How many Wal-Mart stores are there in the United States?
   a. Nearly 1,100.
   b. Exactly 1,100.
   c. More than 1,100.

3. What material was used to make the benches inside the Wal-Mart store?
    a. Steel.
    b. Synthetic wood.
    c. Recycled plastic.

4. What is done with the waste water that is collected in the holding pond?
    a. It is used to wash the building.
    b. It is used to water the shrubs and grasses.
    c. It is poured into a nearby river.

5. Which of the following statements is true about the new Wal-Mart store?
    a. It has been designed so that it can be recycled.
    b. It used to be a factory.
    c. It probably cost less to build than the other Wal-Mart stores.

CHECKING WHAT YOU HEAR

Watch the last part of the video. What do Susan Maxman and Barry Serafin say about "green architecture"? Check (✓) the statements that agree with what they say.

1:51:09–
1:51:29

1. ❏ It's a style.
2. ❏ It's a fad.
3. ❏ It's our survival as a species and as a society.
4. ❏ It's still a novelty.
5 ❏ It can make use of a wide array of new technologies.
6. ❏ It's becoming more available.
7. ❏ It's getting to be more cost effective.
8. ❏ Architects can easily ignore it.

AFTER YOU WATCH

LANGUAGE POINT: COMPOUND WORDS

Compound words are nouns, verbs, adjectives, or adverbs that are composed of *two or more words or parts of words*. Two examples of compound words that are used on the video are *skylights* (windows in a roof or ceiling) and *blue jeans* (casual trousers made of denim cloth).

The words in the boxes on the following page can be used to make compound words that are used on the video. Match each word in box A with a word in box B to make a compound word that fits one of the definitions. The first one has been done for you.

A		B	
day	main	back	~~place~~
head	out	light	pond
holding	pay	grow	quarters
lumber	~~work~~	stream	yard

1. The place where you work is your __*workplace*__ .
2. The main offices of an organization are its _____ .
3. The light during the day, or the time of day when it is light, is called

   _____ .

4. Money you get in return on an investment is sometimes called a

   _____ .

5. Freshly cut wood is stored and sold in a _____ .
6. A major trend or a general direction in which ideas are flowing is called

   the _____ .

7. If you _____ something, you get too big to use it.
8. An artificially created area that is used to hold water is called a

   _____ .

VOCABULARY CHECK

The following sentences are from the video. What do the italicized words
mean? Circle the correct answer.

1. Instead of building a new structure, architects recycled an old one,
   saving hundreds of tons of steel and concrete and 9,000 tons of
   masonry.
   a. woodwork   c. stonework
   b. paint   d. plastic

2. We have a payback in energy alone — $100,000 a year — that will very
   quickly *recoup* any margin of additional investment.
   a. use up   c. destroy
   b. make up for   d. borrow from

3. The carpeting is natural wool with no dye or glue to *give off*
   chemical fumes.
   a. damage or destroy   c. remove or take away
   b. take in or absorb   d. emit or discharge

4. Finding environmental building materials can be difficult, but suppliers are beginning to *sprout up*.
   a. appear
   b. disappear
   c. raise their prices
   d. lower their prices

5. The new ideas are already beginning to *show up* in surprising places.
   a. make people laugh
   b. be used as decoration
   c. vanish
   d. occur

6. Outside, waste water is collected in a holding pond, treated, and then used to water the native shrubs and grasses used for *landscaping*.
   a. planted gardens
   b. cleared the land
   c. construction on land
   d. food

7. For *advocates* like Susan Maxman, president of the American Institute of Architects, "green architecture" is already beyond the experimental stage.
   a. attackers
   b. supporters
   c. experimenters
   d. designers

8. With a wide array of new technologies, it ("green architecture") is becoming more available, more *cost effective*, and harder for architects and the companies that hire them to ignore.
   a. precious
   b. expensive
   c. economical.
   d. wasteful

DISCUSSION

Work in groups. Discuss your answers to the following questions.

1. What did you learn from the video?

2. In your opinion, which of the two environmentally friendly buildings featured on the video (the National Audubon Society headquarters and the Wal-Mart store) is more attractive or interesting? Explain your answer.

3. Describe the school or institution where you study. What is one thing that could be done to make it more energy efficient?

4. Do you feel it is really important to try to take care of our environment and to conserve energy? Why or why not?

ROLE PLAY

Work in groups of four. One student will play the role of the father, the second the role of the mother, the third the role of the son, and the fourth the role of the daughter. Read the situation and the role descriptions and decide who will play each role. After a ten-minute preparation, begin the role play.

THE SITUATION: **A Family Discussion**

A middle-income family is having a discussion about their plans to build a house on land that the children have inherited from their grandfather. The family is designing the house themselves. The first step in the planning process is to limit the number of features in order to stay within the budget—both during construction and afterward, when they will have to maintain their house.

ROLE DESCRIPTION: **Father**

You are a forty-five year old electrical engineer. Years of experience have taught you that there must be a best way of building a home to make it energy efficient. You want to use solar energy as much as possible, and you would like to keep the family's electric bills down by installing solar water heaters. Because your experience with building solar heating systems is limited, you need to keep things as simple as possible. During the discussion, you often offer advice about the work involved in each of the projects that the other members of your family mention.

ROLE DESCRIPTION: **Mother**

You enjoy cooking and would like your kitchen to be equipped with the most modern and time-saving appliances. You are willing to experiment with a solar oven, and you would also like a solar greenhouse in which you could grow all types of vegetables and herbs.

ROLE DESCRIPTION: **Son**

You are a seventeen-year-old who plays the drums in a band at dances and parties. You would like a solar-heated, glassed-in porch where you can hold band-practice with your friends — even in the dead of winter — and not have the noise bother the rest of your family.

ROLE DESCRIPTION: **Daughter**

You are a fifteen-year-old who is very social. You would like to be able to invite your friends and teammates on the school swimming team to the house to practice in an in-door pool all year round. For you to do this, the pool would need to be heated by the sun.

READING

The reading below describes a single-family home built thirty to forty years ago. Like many older homes in the United States, it has some problems. Read the description to find out about the house and the problems it represents to the family who live in it. Then answer the questions.

THIS OLD HOUSE

Living room (14' X 16')

Two large windows allow the daylight to filter into the room to provide an open, airy look. At the same time, the single window panes allow much of the heat to escapte in the winter. During the summer months, the air conditioning has to work hard to make the room cool enough to be comfortable.

Bathroom (6' X 9')

All four members of the family take long showers daily in the one bathroom. After only two people have showered, the heated water in the water tank always runs out.

Kitchen (10' X 12')

The refrigerator needs to be defrosted every month. The insulation in the oven door has deteriorated, so much of the heat from the oven seeps out into the kitchen area. This is very comfortable during the winter, but suffocatingly hot in the summer.

2 Bedrooms (12' X 13' and 14' X 15')

The smaller of the two bedrooms has no carpet because one of the sons who sleeps in it is allergic to wool. As a result, the floor is very cold during the winter. The larger bedroom has no window curtains because the mother likes a simple style instead of heavy drapes for the window treatment.

1. Which room in the house is the largest?

   _____

2. What could be changed in the living room to make it more energy efficient?

   _____

3 What could be done to solve the problem with the bathroom?

   _____

4. What can you conclude from the fact that the refrigerator needs to be defrosted once a month?

_____

5. What do you think the phrase "window treatment" means?

_____

6. How many problems are there in the house? Describe them.

_____

WRITING

Complete one of the following activities.

1. You work for the National Audubon Society in New York and have been asked to prepare a press release for a local newspaper. Write 150–200 words about the society's new headquarters — the reasons for recycling an old building instead of constructing a new one, the environmentally friendly features of the design, and the benefits it has for the National Audubon Society and the environment.

2. You live in Lawrence, Kansas. You have just visited the new Wal-Mart store. Write a letter to the editor of a local newspaper, giving your opinion of the new store and whether or not you feel other retailers should follow the example of Wal-Mart.

What Are the Differences Between Men and Women?

from *Nightline*, July 31, 1991

Forest Sawyer: It took the Gulf war to make us notice just how much has changed between the sexes. Day after day of the crisis, we saw men and women working side by side in the field. We saw fathers and mothers separated from their families, putting themselves in harm's way.

The U.S. Senate noticed, too. Today it joined the House in voting to allow female military pilots to fly combat missions, which stirs up an old hornet's nest. In the old days, it was politically fashionable to say women are the gentler sex, and then it was fashionable to say the only real difference between little girls and little boys is the plumbing.

And now? Well, now we're confused. So tonight, let's bring you up to date. Science's latest picture of what makes men men and women women. First, some background from Harvard scientist and ABC correspondent Michael Guillen.

Michael Guillen, *ABC News*: (*voice-over*) Scientifically speaking, Greg and Terry are members of the species Homo Sapiens, the personification of millions of years of evolution. Politically speaking, though, they're pilots in the U.S. Air Force who happen to be members of the opposite sex, the personification of an age-old debate that heated up again today in the Senate.

Sen. William Roth, *(R) Delaware*: Women have proven themselves. The documentation is clear and well-documented. The best arguments are performance, experience, and aptitude. And women military pilots have come through with flying colors on all three accounts.

Sen. John McCain, *(R) Arizona*: I think that it's important to recognize again this is a national defense issue, it's not a women's right issue—rights issue.

Michael Guillen: Today's debate was carried out in a political arena, but it raises questions that are being studied in a scientific arena.

Do women have what it takes to be combat soldiers? Are males the superior sex or are females? What science is discovering might make politicians think differently about the differences between the two sexes.

(*voice-over*) Globally, there are as many as 50 million different kinds of plants and animals. Some, like snails, have only one sex, and some, like honeybees, have what amount to three different sexes. The vast majority of plants and animals, though, have two sexes.

Deborah Charlesworth, *Evolutionary Biologist*: If you have two sexes, you avoid inbreeding. If you're just male or just female. you can't self fertilize, you can't have inbred progeny.

Michael Guillen: (*voice-over*) Among elephant seals, the two sexes are very different from one another, in size, weight, and temperament. On the other extreme is the wolf, where the two sexes are very similar. They have roughly the same body weight and both sexes hunt and take care of the young. Increasingly, scientists are finding that human beings are more like wolves than like elephant seals. In fact, they've discovered that in the early states of life, the two sexes are virtually indistinguishable.

Melissa Hines, *Neurobiologist*: Well essentially, until about week six of gestation, they're identical, the male and the female fetus. They have the same sex organs that will eventually develop in the female into an ovary and in the male into testes, the same vital organs, the same reproductive structures.

Michael Guillen: (*voice-over*) After six weeks, some fetuses suddenly begin to produce testosterone. At that point, they clearly become males.

Terrence Deacon, *Neurobiologist*: Those hormones begin this inexorable process of altering the plumbing, of altering the structure of the body. If they weren't there, we would develop into females, all of us.

Michael Guillen: (*voice-over*) At birth there are obvious sexual differences, of course. But beyond that, scientists are realizing something quite remarkable.

Melissa Hines: At birth, the male and female infants are very much the same, despite these physical differences, and even in adulthood, although we see some differences between the sexes in behavior, by and large, we're similar.

Terrence Deacon: That doesn't mean there aren't sex differences, and it doesn't mean there aren't consistent sex differences.

Michael Guillen: *(voice-over)* In adulthood, the average male is bigger than the average female; bigger heart, bigger lungs, bigger body, bigger muscles. On the flip side, the average female has better senses; better sight, better taste, better smell, better touch, better hearing. When it comes to the brain, there are a few differences, but outwardly things look pretty equal, even to the expert eye.

Melissa Hines: If you were to put two brains in front of me, I would not be able to say for certain whether one was a male and the other female, or whether they were both males or both females.

Michael Guillen: Females live longer, by seven years, even though males start out ahead of the game: 105 males are born for every 100 females. But then they start to die off. By middle age, it's down to 98 males for every 100 females, and by old age only 72 males are left for every 100 females.

A soldier is supposed to be aggressive, and on that score there is a big sexual difference. Every year, about 15,000 males commit murder, compared to 2,000 females. Is aggression in humans caused by male testosterone? Well here the jury is still out.

(On camera) When it comes to animal studies, things are a little more clear-cut. When you inject female rats with testosterone, for example, they do become more aggressive. But since rats and humans are so different genetically, drawing any conclusions from this study raises more questions than it answers.

(voice-over) A soldier also has to be disciplined, and on that score the testosterone can be a liability. According to a recent study of Vietnam veterans, soldiers with the highest testosterone levels were most likely to go AWOL or end up in the brig.

Terrence Deacon: Probably it's the case—the individual variety. All kinds of behaviors that we measure swamps the sex differences. We will never do a good job if we use sex differences as our major measure of the difference between people, because there is so much of this variety.

Michael Guillen: *(voice-over)* Take height, for example. The average woman is about 64 inches tall, the average man five inches taller. But there are many, men and women, who fall above and below the averages.

Melissa Hines: I think that science has to say is that men and women run the full gamut of behavior and of brain structure.

Michael Guillen: *(voice-over)* Increasingly, scientists are finding that these kinds of variations exist for all traits, including strength, endurance and aggression, so that in the scientific arena, at least the old idea that one sex is absolutely superior to the other is proving to be just a lot of hot air. I'm Michael Guillen for *Nightline*, in New York.

TRANSCRIPT 2

You Can Work It Out

from *20/20*, November, 29, 1991

Hugh Downs: If you're married, how is your marriage going and if it's not going well, do you care enough to save it? A decade ago, couples seemed more willing to head right for the divorce court, but now many marriage counselors report more couples are trying to work it out. But exactly how do you do that and isn't every marriage different?

(voice-over) Stone Phillips has a couple whose marriage was in trouble for a very specific reason and yet, their problem is one that affects millions of marriages. In fact, it could be spoiling your relationship and you might not even know it.

Stone Phillips, ABC News: *(voice-over)* Perhaps the single most perplexing question when it comes to marriage is why two people, starting out with the greatest of hope and the best of intentions, so often wind up disillusioned, slipping into roles and behavior that neither partner likes but just can't seem to avoid, as if marriage itself somehow turned them into people who no longer bring out the best in one another. This is the story of Bob and Judy Muller of Brunswick, Maine and their struggle to save their marriage of six years. Judy recently became a family therapist. Bob does marketing for a local map-making company. It is the second marriage for both and they're hoping to have children soon. Currently, they're building a house and doing most of the work on it themselves. The conflicts they are facing now were just unimaginable on the night they met eight years ago on a blind date.

(Interviewing) When did you fall in love?

Judy Muller: That first night.

Stone Phillips: Smitten?

Judy Muller: Mmm . . .

Stone Phillips: From the beginning?

Judy Muller: Yup

Bob Muller: Ready?

Judy Muller: Yup..

Stone Phillips: *(voice-over)* And it wasn't only a romantic attraction. The moment they laid eyes on each other, they saw qualities they really liked.

Judy Muller: I got dressed up and he came to the door and s   that I looked really nice and asked — and he was in jeans. An so, he asked me whether he w   ed me for him to change and I said, "Yes."

Stone Phillips: What did you think about that? Did that take you back a little bit?

Bob Muller: No! I immediately said, "Now, that's a lady I can relate to." She said exactly what she wanted. It wasn't wishy-washy or anything like that — and I said, "I can handle that."

Stone Phillips: As you got to know Bob better, what kind of husband did you expect he would b

Judy Muller: I thought that he would be open and generous and fair.

Stone Phillips: *(voice-over)* Her independence, his open-mindedness — these were the qualities that attracted them, but not long after they were married and had moved in together, their behavior toward one another began to change.

Judy Muller: He seemed less open and more controlling and more sort of driven to do what he wanted to do.

Stone Phillips: And it was easier for you to fall back on—

Judy Muller: Letting him take care of me, letting him make decisions.

Jo-Ann Krestan, *Marriage and Family Therapist:* Two people fall in love, they get married and they instantly go into kind of almost roles that they've been hypnotized into playing.

Stone Phillips: *(voice-over)* Jo-Ann Krestan is a marriage therapist who has counseled thousands of couples, including Bob and Judy Muller. Krestan says it is an almost universal phenomenon husbands and wives adopting what she calls "gender roles" from their own mothers and fathers and it often leads to trouble in marriage.

Judy Muller: My father makes a lot of the decisions and my mother goes along with it and she doesn't say anything. And I think that I kind of fell into that role, too.

Bob Muller: And I was brought up, you know, where my father had all the power. You know, he was the breadwinner. My (my) mother didn't (didn't) work. You know, she took care of the family and that was my training.

Stone Phillips: *(voice-over)* Judy also found herself imitating her mother's way of being the emotional support system for the family. And so even though Judy was financially independent, with a good job of her own when they first met, there was no question for either of them who would be the one to pick up roots and move. Judy joined Bob in Denver where he was working. When he found a better job in Washington, DC. Judy moved again. By the time Bob moved them back to Maine for a new job and a new dream to build this house, Judy had deferred her needs to Bob's plans so many times that she had lost all sense of herself.

Judy Muller: And I was feeling frustrated and lost. The relationship was changed. It was changing and things just weren't working and I didn't like it. I didn't like how I was feeling.

Stone Phillips: *(voice-over)* Jo-Ann Krestan says that Judy's depression is typical of what happens when people get locked into a gender role. A book she co-authored explains how women get married and then become too good for their own good, deferring to their husbands and taking care of them emotionally so much that women stop making their own decisions and taking responsibility for their own lives. Men pay a price, too, because when women do all the emotional work for the marriage, men tend to withdraw and grow out of touch with themselves.

(interviewing) So that kind of structure becomes embedded in the marriage relationship?

Jo-Ann Krestan: And it lets everybody off the hook in terms of taking the responsibility for figuring out their own life goals. And I think we merge our identities in marriage to such a degree that people really lose a sense of themselves and of responsibility for themselves. And when they lose who they are, they really lose a sense of joy and of aliveness. A kind of deadness creeps in.

Stone Phillips: *(voice-over)* The joy had gone out of Bob and Judy's marriage. Two years had passed since their last move and they were still living in their small temporary trailer while the dream house grew bigger and bigger. They were now bickering and arguing. Finally, the anger erupted over the project they were locked into together, the building of the new house.

And listening to their arguments, you can see how many of the troubles they've gotten into come from their gender roles — Bob making the decisions, Judy deferring.

Bob Muller: All of a sudden now it was my design and my plan, and it's too expensive and it's all my fault. There were two people here during this whole process. Maybe I resent her for not speaking up earlier in the stage.

Judy Muller: I kept trying to say no, but I just couldn't do it. I couldn't do it strong enough.

Stone Phillips: *(voice-over)* Feeling frustrated and utterly discouraged, Judy and Bob went for help.

(Interviewing) Bob and Judy Muller, what - what's the problem dynamics in this marriage?

Jo-Ann Krestan: When I first saw Bob and Judy, Bob was yelling at Judy for buying a $30 blouse and he had just bought a $2,000 tractor. And it was that assumption—that he earned the money, he had the right to decide how they spent the money. His career came first, his job came first— and that root assumption really set the ground for a lot of the problems that were going to come down.

Stone Phillips: *(voice-over)* It was only in therapy that Judy began to see the extent to which she was deferring her whole life to Bob and that to stop doing it, she would have to focus on herself and not feel guilty about it. She finished her graduate work and went to work in a community clinic. Judy was developing a new sense of herself to replace the old gender role that had served her so poorly.

The impact of that change hit home on the day Bob walked in from work with a major announcement. He was going to change jobs again. He had been offered a job out of state and would commute back to Maine on weekends. Well, this time, Judy stood up for herself. She told Bob she didn't want an absentee husband. If he took the new job, she would leave him.

Judy Muller: I got tired of giving up my life all the time for his dreams. And I wasn't willing to do that anymore.

Stone Phillips: *(voice-over)* But even though Judy had finally stood her ground, three months had gone by and Bob still hadn't made a decision about whether he would take the new job.

Jo-Ann Krestan: What's going on with this? Is there any movement?

Judy Muller: I don't feel like there's a whole lot of movement. I feel like we're — I feel like it's stuck and I feel like I'm not supposed to say anything and I'm not supposed to force the issue because it's not my responsibility.

Jo-Ann Krestan: To force the issue? It's an issue of where you're going to live.

Judy Muller: But it's not my — I'm not the one who is going to change jobs or not. He is. He's the one who's going to figure out what he's going to do.

Jo-Ann Krestan: Wait a minute, wait a minute. Let's get clear on whose — this is an issue that affects your life.

Judy Muller: Right

Stone Phillips: *(voice-over)* Jo-Ann Krestan helps Judy stick to her guns.

Jo-Ann Krestan: You never give Bob a timetable.

Judy Muller: Because I'm afraid of the answer.

Jo-Ann Krestan: Yes exactly, and that Judy, that's exactly why couples don't come to clean positions with one another in which they really get to their bottom line because they're afraid of the reaction. What would happen if you just said, "I want closure"? And then, you'd know what you were dealing with. What timetable would be comfortable for you for closure? Then you'd know.

Judy Muller: It's hard for me to answer that because it's hard for me to ask for that.

Jo-Ann Krestan: I know. I want to help you. I've always wanted to help you ask. I have always wanted to somehow say to you, Judy that as a woman, I know how hard it is to ask. I know how hard it is to feel like I have the right, like I'm entitled.

Stone Phillips: You were, you were moved to tears in the therapy session when Jo-Ann told you you've got a right, you're entitled to your opinion to be assertive, to tell him what you think, to take a position. How are you feeling?

Judy Muller: Scared — scared that he, that he would leave me, that it wouldn't work out, that it wouldn't be okay if I did that.

Stone Phillips: Why did you encourage Judy to set that deadline?

Jo-Ann Krestan: I wanted to support her in taking a position for herself which was that the anxiety was simply getting intolerable

Judy Muller: I need to have some closure.

Jo-Ann Krestan: By?

Judy Muller: By February 14th because it's too hard to keep going around and over and over.

Stone Phillips: *(voice-over)* Now, it was Bob's turn because, according to Jo-Ann Krestan, when one person changes in marriage therapy, the other person will soon either leave the marriage or change, too.

Jo-Ann Krestan: It's just like taking a mobile and removing one little piece. For a while, it's in disequilibrium and everything swings wildly until it finds a new balance.

Stone Phillips: *(voice-over)* Bob faced a choice. In order to stay in his marriage, he had to stop constantly moving in search of new jobs. The decisions he had made, which he told himself were for his marriage, were in fact putting his marriage in jeopardy, so what was actually motivating him, making him drive himself so hard?

Jo-Ann Krestan: This is right back to where we were last time. What is driving you?

Stone Phillips: *(voice-over)* Just as earlier in the therapy session, Jo-Ann Krestan pushed Judy to do her own decision-making and not defer to Bob. She now pushes Bob to do his own emotional work and not depend, the way he always has, on Judy.

Bob Muller: I mean, I feel if I knew the answer to that — maybe that's at the core of the way I deal with a lot of things.

Jo-Ann Krestan: What's the most personal conversation you ever had with your father?

Bob Muller: I don't remember a very personal conversation. He was in the Navy. He was gone six months at a time on deployment, so we wouldn't see him for a whole six months.

Stone Phillips: *(voice-over)* This new information confirmed Jo-Ann Krestan's hunch that in proposing to take a job that required him to be away from home, Bob was repeating a pattern he had learned form his father.

Jo-Ann Krestan: So those patterns sort of filter down into marriage.

Stone Phillips: He's doing what his father did.

Jo-Ann Krestan: Absolutely and quite unconsciously, until, as part of my job, I help make that conscious and flush that out.

Do you have a memory of his coming home at any point? Do you remember going up to him and wanting to talk to him? Do you ever remember trying to get his attention?

Bob Muller: No. Maybe during football games or looking—

Jo-Ann Krestan: Did he come to your football games?

Bob Muller: He tried to. I remember looking up to see if he was in the stands.

Jo-Ann Krestan: And so you played for him. You played for him. What's hurting?

Bob Muller: I don't know.

Jo-Ann Krestan: It's okay. I want to understand this, Bob. Judy, take my chair.

Stone Phillips: *(voice-over)* Jo-Ann asks Judy to sit apart. Bob has to come to terms with his emotions himself.

Bob Muller: I'm still living his life and not mine. I don't know if I'm living the life I want to live. Either I do it for this reason or this reason or them or her or him, but what am I doing for myself?

Stone Phillips: *(voice-over)* Jo-Ann Krestan offered Bob a suggestion — why not try to talk with the father he had never really gotten to know? Though they lived only 10 miles apart, they seldom saw each other. Jo-Ann thought if Bob could find out more about his father's life, it might help him better understand himself. So, here, for the first time in his life, Bob is asking his father about his own mom and dad.

Bob Muller: I have a big blank there 'cause I never knew my grandfather.

Jack Muller: Well, my mother and father got divorced at an early age and I was just starting high school and I never did see my father after that. Back then, divorce wasn't a very common thing. You know, that, that to me was a bad thing as far as that goes. It was like —

Bob Muller: Well, you must have been very mad at him or pissed off or —

Jack Muller: Oh yes, I was. The further I could get away from it, the better off I thought I would be, but as I say, that's — hell, I went in the Navy when I was 17. It was something to do — get out, get a new life.

Stone Phillips: *(voice-over)* Bob could now understand where he'd learned the pattern of leaving home to get away from problems.

Bob Muller: You've carried over what you learned from your father and it's carrying over to me.

Stone Phillips: *(voice-over)* But there was another thing he was beginning to suspect as a result of this conversation with his father. Perhaps the reason he kept driving himself to more impressive jobs and a bigger and bigger house was to get the attention of the father who had been cut off from him in childhood. If he could finally make an emotional connection with his father, maybe he could stop driving himself.

Bob Muller: I love you very much, but I don't feel close to you. We talk about all this mechanical stuff, but I never seem to hear the loving side of things and I know we're capable of showing that.

Jack Muller: Oh, yeah, I agree. you're a little different than I am in that respect. You want to hear it, right?

Bob Muller: Yes.

Jack Muller: And unfortunately, I —

Bob Muller: I want to hear it and feel it.

Jack Muller: I never was really brought up that way. So, I don't know what to tell you.

Bob Muller: Can you tell me you love me?

Jack Muller: Of course. I told you I love you. Yes, I love you. I sure do. Whatever gave you that particular feeling I didn't love you?

Stone Phillips: *(voice-over)* Bob has been able to begin an important journey of discovery today because he made the decision to make his emotional life his own responsibility. He has also recently made the decision to stay put in Maine.

Bob Muller: Another day's work. How many more we got?

Judy Muller: A couple.

Stone Phillips: *(voice-over)* If there is one lesson to be learned in Bob and Judy's story, it is that in each partner taking responsibility for all aspects of their lives, they begin to grow again as individuals and as a couple in marriage.

Bob Mueller: What's for dinner?

Judy Mueller: Pizza

TRANSCRIPT 3

Children of Divorce

from *20/20*, March 4, 1982

Hugh Downs: Listen to these children.

Girl 1: Well, I just live with my mother. Um, I don't know where my father is.

Girl 2: My parents have joint custody, and half the week I live with my mother, and the other half I

126

live with my dad and my stepmother.

Boy 1: I switch off every day — every other day, mother...father.

Girl 3: I live with my mother, not my father, and I don't know where he is.

Girl 4: I live with my father and my stepmother. I don't know where my mother is.

Boy 2: I live with my mother, and — that's it, my mother and my pets.

Hugh Downs: These children are part of one of the fastest-growing population groups in the United States — the children of divorce, thirteen million of them in this country right now. And sadly, because divorcing adults are so often absorbed with their own pain, the children's voices are seldom heard. But that's changing — now there is new attention to the children, new studies on the effects of divorce from their point of view. And here with a report is Bob Brown. Bob?

Bob Brown: Hugh, divorce is a topic almost everyone has strong feelings about and can debate endlessly, especially if they have been through one. But we're not going to hear much from parents in this report, or about who's right or who's wrong. We're going to see and hear children — who often couldn't care less about the issues adults are familiar with, but who simply must try to deal with the shock when a family breaks up.

(voice-over) Like a scattered puzzle whose pieces can't be forced back into a pattern, the family life that three-year-old Michael has known is ending. His parents are divorcing. They have gone to a mediator to try to work out a settlement between themselves prior to their court appearance.

Secretary: We'll be talking with your mommy and daddy, and we'll go back into the office, all right?

Michael: *[crying]* No, I want to stay in here.

Father: No, you're not going to stay here. Now, Michael, they're going to bring you back into this room —

Bob Brown: *(voice-over)* Michael is afraid of being separated from them, but he will be visited, alone, by a counselor who wants to watch him play.

Therapist: The mama and the papa are together, huh?

Bob Brown: *(voice-over)* How Michael plays reflects how he feels. He takes figures that represent a mother and a father and a child, and holds them tightly together, as he wishes his own family to be.

Therapist: They don't leave each other, do they? They go everywhere together. I think that's how you like it to be in your family, too.

Bob Brown: *(voice-over)* Then, through the same toy figures, he expresses his fears about the meaning of divorce.

Therapist: What happened?

Michael: They got shot.

Therapist: They got shot! Oh, my gosh! And who is the bad guy?

Michael: The hunter.

Bob Brown: *(voice-over)* Through the tools of children, the numbers of divorces become more than detached statistics. They become simple Crayola drawings of worlds falling apart, of children who stand on top of rainbows and wonder which way to go, as their parents slide down into separate homes. And of houses that are swamped by an ocean and lost in the turmoil. Over the last twenty years, the divorce rate has increased by almost 250 percent. Today, there are thirteen million children below the age of eighteen whose parents are divorced. But until the last few years, there were virtually no long-range studies to chart how children respond to divorce. One of the first to do so, published in 1980, was co-authored by Dr. Joan Kelly.

Dr. Joan Kelly, *Psychologist*: Since 1972 there have been one million new children affected each year by divorce — that's a lot of youngsters. And it is in retrospect, to me, sort of embarrassing that we in the mental health profession really didn't think through issues of divorce.

Bob Brown: *(voice-over)* Professional researchers weren't the only ones bothered by the lack of attention paid to children of divorce. Because they felt their own reactions were either misunderstood or ignored, a group of eleven to fourteen-year-old students, here at the Fayerweather School in Cambridge, Massachusetts, wrote *The Kids' Book of Divorce*. In it they included advice to other children. "All kids need to know these things," they wrote. "They need to know that divorcing parents don't love each other, but they still love their kids. They need to know that life may be very hard before their parents reach a settlement. They need to know where they are going to live." Among the twenty authors of the book, all in high school now, are Sarah Steele, Louis Crosier, Matthew Allison, and Heather Murphy.

Heather Murphy: I came from a very rural town, where not many people were divorced, and if they were, they didn't talk about it. And so I didn't really understand the subject. I didn't understand many things about life. And so by writing the book and being able to express my feelings — which is something I had never been able to do before — I think I came out of myself and I really found out who I was.

[clip from *Children of Divorce*]

Bob Brown: (*voice over*) The old concepts of who the children of divorce were and what might become of them were reflected in this 1927 film titled, *Children of Divorce*. A girl is dropped off by her mother at an American divorce colony in Paris. She's not only lonely and frightened, but as the film shows later in a disapproving view, she will grow up without morals or love — a notion that was often believed but had no basis in fact.

[clip from *Shoot the Moon*]

Girl: I hate Daddy.

Bob Brown: (*voice-over*) Today, in films such as *Kramer vs. Kramer, Only When I Laugh*, and this recent film, called *Shoot the Moon*, there is a more real and sympathetic view of the troubles of children, based on the research that's being done, and concern for trying to explain a shattering experience in ways that speak directly to a child's questions.

[clip from *Shoot the Moon*]

Girl: Why did Daddy leave us?

Woman: Well, I don't think he left you. I think he left me.

Nicky: First it felt totally negative that my parents weren't going to be living together.

Bob Brown: (*voice-over*) Nicky was six when his parents divorced.

Nicky: And, but then, I mean, they kept telling me that it was all probably better and everything.

Bob Brown: Though it may seem better to the adults, it seldom seems that way initially to the children. Divorce is painful at any age for a child. What changes with age isn't the pain, but the way children react to it. And that's one of the most important findings researchers have made — identifying the ways in which children deal with divorce, depending on their age.
(*voice-over*): Preschool children, for instance, are most likely to blame themselves for the divorce. Some become more aggressive, others simply withdraw.

Therapist: [*through hand puppet*] Good-bye, I'm going to miss you.

Bob Brown: (*voice-over*) These preschoolers are in a play therapy group for children of divorce at California's San Fernando Valley Child Guidance Clinic. Because younger children are less able to talk about how they feel, counselors use hand puppets as a device to bring emotions out into the open.

Therapist: And it makes me sad, and you know what else?

2nd Therapist: What?

Therapist: It makes me think about the time I had to say good-bye to Daddy.

2nd Therapist: Yeah. That was so sad.

Therapist: That was really hard to do.

Bob Brown: (*voice-over*) Researchers are finding that after a parent leaves home, the normal course of every other weekend visitations falls far short of meeting the needs of children.

Dr. Kelly: Certainly one of the major factors that determines how well children do after a divorce, is the extent to which children have contact with both parents. We have used the term 'single parent family' in this society, and I think it's an unfortunate choice of words because the child really remains committed to the notion of the two-parent family. And the single-parent family concept somehow implies that a child doesn't have two parents after a divorce.

Bob Brown: (*voice-over*) While preschoolers are frightened and bewildered by the absence of a parent, the response begins to differ in children around the ages of five or six. They understand more, and are also more likely to retreat into grief and depression. They often become disorganized in their lives, while they try to remain composed. And around the age of eight or nine, there is another shift in response. Becky Arnold is a nine-year-old whose parents are separating. They live in the suburbs of Fort Lauderdale, Florida. Becky's reactions are typical of those in the nine to twelve-year-old age group: she is able to talk much more openly about her fears and her anger at the divorce.

Becky Arnold: My mother told me in the middle of the night.

Bob Brown: What were you feeling inside?

Becky Arnold: Sad, and mad, that my dad ran off without even telling me. I was afraid I would never see him, and he would always stay out, and never come over to see me.

Bob Brown: *(voice-over)* Becky's parents represent the rule rather than the exception. Most parents find it difficult to communicate well with their children about divorce, and they may even seem to be unaware of the depth of their children's pain.

Mr. Arnold: I think it's very difficult for adults to be in touch with their children's feelings, if they are having a very difficult time with their own feelings.

Bob Brown: *(voice-over)* Becky goes to a group in Fort Lauderdale, where she can talk with other children near her age. One of the reasons they can't talk with their parents is that children in this age group are often caught between disputes of divorcing parents.

Becky Arnold: *[at group session]* And my dad, he lives in this house, and my mom wanted to get something that was in there, and she put me through a window. And my dad said, "Did you happen to go in the house down the street?" And I was — I couldn't say it, because my mom said not to tell anybody.

Every time I came home from going out with him, she says, "Is there anything that he said?" And I really tell her all, and I feel like I'm in the middle.

Bob Brown: *(voice-over)* And researchers found another complication that is common to children Becky's age. Sometimes, even years after a divorce, they are still absorbed in a wish that their parents will reunite. Some children devise schemes to try to get their parents back together. Rick ran away, and spent the night with friends.

Therapist: When you ran away, what did you think might happen?

Rick: They'd both get together and search for me, and I thought they might realize they should be together, and they —

Therapist: Okay, I follow you.

Becky Arnold: My dad was back for a while, for like, he spent the day with us. And I thought that if he spent the day with us, then he would come back. And it turned out at the end of the day, they ended up fighting.

Mr. Arnold: I believe she is still expecting me to come back. And I've told her that I'm not coming back, and she really doesn't want to hear that, so I don't go into that with her.

Bob Brown: What would you like to see happen now?

Becky Arnold: My parents get back together, or at least be friends.

Bob Brown: Do you think that might happen?

Becky Arnold: Maybe.

Bob Brown: *(voice-over)* Most children of divorce must in time leave the wish behind them, and it is becoming a common disillusionment. Sixty percent of all divorcing families have children in Becky's age group or younger. But even when adolescents are involved — thirteen to eighteen-year-old who are more mature — the impact of divorce is still great.

Isolina Ricci, *Family Counselor:* Adolescence is a time when you're really expecting everything in your lief that isn't totally revolving around you to stay still, so you can get on with this business of developing who you are, and —

Bob Brown: *(voice-over)* Family Counselor, Isolina Ricci.

Isolina Ricci: This is not the time to have parents who are suddenly so self-absorbed, and so involved in sometimes sheer survival, so that the focus isn't on you any more.

Bob Brown: *(voice-over)* Not only are teenagers often caught between divorcing parents, as are younger children, but they're also more aware that they're being used for emotional support by their parents at a crucial time in their own emotional lives. Debbie Duffy is seventeen and lives with her mother.

Debbie Duffy: She used to call me her rock, like she'd — hysterical fits and everything — and cry, and I'd be there. I mean, I went through it, too, but, you know, we'd both hang on to each other and help each other out.

Bob Brown: Did you feel as if she needed you more than you needed her at the time?

Debbie Duffy: Yeah, yeah.

Girl 1: You see them as people who make mistakes and who have feelings, and you also see yourself as somebody who has potential to be something, to make your own decisions.

Girl 2: You know, you have to do more, you have more responsibilities —

Girl 1: Exactly.

Girl 2: — which makes you grow up.

Bob Brown: *(voice-over)* Alexandra Urbanowski is seventeen. She lives with her father, Frank Urbanowski, a Massachusetts Institute of Technology publishing executive. Her friend, Ariadne Horstman, also lives with them. Her parents are also divorced. The extra responsibilities both girls assume have several dimensions and one of the most important is learning to deal with their parents' dating, at a time when they're beginning to explore their own feelings about it.

Frank Urbanowski: So I'm probably not going to see a lot of you over the weekend.

Alexandra: Where are you going to be?

Frank Urbanowski: Well, I'll be out.

Alexandra: Where?

Ariadne: Out, where?

Alexandra: When? Who? *[laughs]*

Frank Urbanowski: Different places.

Girl 1: My feelings about my mother dating have changed dramatically since I was young, because when I was really young I was possessive of my mother, and I didn't want her to date other men — well, any men besides my father. And it just upset me when she went out with men. I used to be very obnoxious to the, you know, guy she dated. Like, once I ran a pony cart into one boyfriend's new Camaro.

Girl 2: It can be a real drag, to wake up in the morning and come downstairs, to find somebody standing in your living room in a bathrobe that you've never seen before.

Girl 1: But now, my feelings have changed, because I like it when she dates, because I feel that she has someone to go out with, and she has a companion.

Bob Brown: *(voice-over)* And there's another aspect of divorce on which teenagers seem to have a wiser perspective, but which affects all children of divorce — the weekend Santa.

Girl 1: The parent in the house sets all the rules, and everyone sort of doesn't like that person, and the non-custodial parent is the sweetie who, you know, buys them things and lets them stay up late, and "Oh, your mommy or your daddy did that to you, oh, I wouldn't if you lived with me."

Boy: I try to avoid just having a good time, because that's — it's sort of false. You have to think — say what's on your mind, and if you're not feeling great, then you have to deal with that.

Bob Brown: *(voice-over)* Although a lot is being learned now about children of divorce, there are still many important things we don't know. We don't know how many children never see the parent without custody, and we don't know whether children of divorce are more likely than other children to eventually go through divorce themselves when they are married. We do know that the children wonder about these things, too.

Boy: I think that when I grow up, I'm going to be a lot more careful with relationships that I start, because I don't want to, you know, don't want to have — I don't want to have to go through that myself, or have — if I have kids — have them have to go through it, because I know that I mean, even though it didn't seem to bother me a lot, it's something that I wouldn't want to go through again.

TRANSCRIPT 4

Old Problems: New Cures

from *20/20*, December 9, 1982

Barbara Walters: Right now Hugh Downs is heading for the South Pole, doing a special *20/20* report on the ice covered continent at the bottom of the world. But, before he left, Hugh put together a report on the medical problem than has plagued him and plagued four out of five Americans at some point in their lives. The problem of excruciating, incapacitating back pain. And I'll bet there isn't any one of you who doesn't know someone who suffers from it, if not yourself.

Woman: Terrible pain. You can't move. I can't move at all. The position that I get in when I get my pain is just flat on my back with my legs up over something. I just can't move at all.

Man: Somehow I reach for something, bend the wrong way, and I have a terrific pain that stays with me about a week or two weeks.

Hugh Downs: *(voice-over)* Backache — it affects everyone, from laborers to office workers to weekend gardeners. And as old and widespread a problem as it is, little is known about its causes or its cures. The major problem in back research is that different irritations within the back can create the same pain, and that's what makes diagnosis so difficult. At the bio-engineering laboratories at the University of Illinois, some of the mysteries of the back are now being explored. For example, we know that improper lifting injures many backs, but we don't know why. This experiment shows how holding a 10-pound weight at arm's length, rather than close to the body, stresses the back muscles by fifteen times.

Dr. Judith Walker, *Walker Pain Institute***:** Eighty percent of back pain is caused only by muscle spasm, and that muscle spasm can be alleviated by bed rest first; if that doesn't work, by proper exercises, by changing people's habits — examining the car seat, for example; examining the chair in which they work daily. Those are very common annoyances that aggravate back pain.

Hugh Downs: *(voice-over)* These people have sought relief at New York City's West Side YMCA's Back class. The exercises are based on relaxation, flexibility, and strengthening. The Y reports that eighty percent of their people show improvement during the program. But there are problems other than muscular that affect the back. One of them is commonly referred to as a "slipped disc".

Let's take a look at the back and try to understand what that means. The spine is made up of vertebrae which stack up, along with discs in between them. When we hear the term "slipped disc", most of us think of a hard piece of bone cutting the spinal cord and causing paralysis. Actually the disc is more like a jelly doughnut with a soft center, called the nucleus, and a somewhat harder outside, called the annulus. As we age, the outside of the disc, the annulus, starts to wear and get brittle, just like getting gray hair or wrinkles. When this happens, the nucleus can bulge out, causing a herniation of the outside. Pressure against the nerves and ligaments can irritate them, causing pain. Irritation of the sciatic nerve can cause sciatica — that's the pain down the leg, so familiar to many back sufferers. Sometimes pressure from a disc can cause a weakness in the leg, or even loss of bowel and bladder control. This is a medical emergency and should be treated immediately.

Back and related leg pain can be so severe that some patients beg for a surgical cure. The most common surgical procedure on backs is a discectomy, which is the removal of the bulging or herniated pieces of the disc, relieving the pressure on the nerve. And what is confusing to doctor and patient alike is that not every herniated disc causes pain. And statistics show that most discs eventually heal themselves, even without surgery. Still, the United States has the highest back surgery rate in the world, and over thirty percent of our operations fail. Because of this, leaders in the field are discouraging surgery as a cure.

Dr. Augustus White, *Orthopedic Surgeon-in-Chief, Beth Israel Hospital, Boston***:** We say that if the surgery is suggested for the first time, you certainly always ought to have two opinions; and if a second operation is suggested, you ought to have three opinions; and if a third is suggested, you ought to have four opinions, etc.

Hugh Downs: *(voice-over)* Augustus White, chairman of orthopedic surgery at Beth Israel Hospital in Boston.

Dr. Augustus White: If you have someone who's saying they're going to cure you, don't worry about a thing — I tell my patients that they probably ought to get their coat and go in the other direction.

Hugh Downs: Stuart Belkin, an orthopedic surgeon in Bridgeport, Connecticut, has been involved in back research his entire career.

Dr. Stuart Belkin, *Orthopedic Surgeon*: Almost any time you go to a meeting where spinal surgery is discussed, there's always one lecturer who is lecturing on failed back surgery. And the number one cause of failed back surgery by far is not a technical error, it's not some problem with the patient's tissues — it's choosing the wrong patient.

Hugh Downs: *(voice-over)* The right candidate for back surgery is chosen under the strictest criteria, and represents the following symptoms: severe back and sciatic nerve pain, no relief after several weeks of bed rest, and in addition has positive indications from special X-ray procedures known as myelograms and CAT scans. Even under these strict criteria, back surgery can fail and leave the patient in a far sorrier state than his original condition. Dr. Belkin frankly tells his patients what risks they face from surgery.

Dr. Stuart Belkin: Failure to relieve pain; increased pain; loss of bowel and bladder control; loss of motor control in the lower extremities; loss of sensation; infection; spinal fluid leak; clots in legs; clots in lungs.

Hugh Downs: Leaders in the field are all too aware of this list of horrors, and they are disappointed in the erratic success record of the 200,000 disc operations performed in the United States each year, especially since this study was done, showing that after five years, surgical patients are not significantly better off than non-surgical patients. Now, for these reasons, the trend in back care is to seek out alternatives to surgery. We are going to show you some new methods of relief that are becoming available to severe back sufferers.

(voice-over) One of these is being studied by Dr. Belkin in an FDA research program. It is the use of an enzyme called chymnopapain, which when injected into the disc, dissolves it completely, releasing the pressure on the nerve.

Dr. Stuart Belkin: You can see that milky material, That's his disc. That's how fast it dissolves.

Hugh Downs: *(voice-over)* The drug has been used in Canada, where 10,000 Americans have gone for help, and just recently it was approved for use in the United States. Belkin has first-hand knowledge of the drug. He chose it over surgery when he slipped a disc. For him the treatment was effective, leaving no after-effects and no scars.

John Dunleavy wishes he had known of an alternative to surgery when he hurt his back on his job as a milkman twelve years ago. Since his injury he has had four operations. Increased nerve scarring from each operation has left him in more and more pain.

Hugh Downs: Were you ever promised a cure through surgery, or that you'd be pain-free after an operation?

John Dunleavy: Yes.

Hugh Downs: That right, then?

John Dunleavy: Yeah, said I'd be fine, and I never was. I never was.

Hugh Downs: *(voice-over)* Dunleavy's pain was so bad that he couldn't sit, stand or walk comfortably for any length of time. Sometimes only narcotics could relieve him. We filmed Dunleavy at Beth Israel Hospital in Boston, where he had come to their pain clinic to find some sort of relief from his pain. Dr. Carol Warfield, an anesthesiologist and head of the pain clinic, decided that the chronic burning pain in John Dunleavy's legs suggested an inflamed nerve root. In such a case, Dr. Warfield favors the injection of a steroid into what is called the epidural space at the base of the spine. In sixty percent of the cases chosen for this procedure, it gives relief. But for someone with as many surgeries and as much nerve scarring as Dunleavy has had, the chances are not as good. Dunleavy decided that he had nothing to lose. After the injection of a local anesthetic, Dr. Warfield administers a combination of cortisone, a potent anti-inflammatory drug, and another anesthetic.

Dr. Carol Warfield: It's a good sign, that it's a little uncomfortable.

Man: Just a little longer.

Dr. Carol Warfield: It means it's right on the right spot.

Man: Another three seconds.

Hugh Downs: *(voice-over)* We'll return to Dunleavy to see how the drug works, but first let's look at another patient, with a different problem.

Abby Summersgill: I woke up in the middle of the night, and my back hurt.

Hugh Downs: Abby Summersgill, lawyer and mother of two sons found her very active life suddenly curtailed.

Abby Summersgill: *(voice-over)* I felt most comfortable in a position like that, crouched over, and of course you can't walk around like that.

Hugh Downs: *(voice-over)* Advanced technology available at Beth Israel made proper diagnosis possible. A series of special X-rays revealed a problem known as facet joint syndrome. The facet joint allows us to bend forward and back. With the natural wearing of our discs, nerves within the joints become sometimes irritated, causing pain such as Abby Summersgill experienced. This problem is often misdiagnosed, and could have led to unnecessary surgery had it not been revealed by the specially angled X-rays at Beth Israel. The treatment is to inject steroids into the facet joint. This is done under fluoroscopy, a special type of X-ray. The fluoroscope is turned on only to see the position of the needle — that's to minimize exposure to the radiation. The fluoroscope can move 180 degrees around the patient, showing the doctor the needle's position from different perspectives. Once the needle is properly placed, the steroid is introduced.

Abby Summersgill: I hope it works.

Hugh Downs: *(voice-over)* It would be three weeks before we returned to see Abby Summersgill's results. In the meantime, we went to the Walker Pain Institute in Westwood, California, where new research by Dr. Judith Walker, an anesthesiologist and neurophysiologist, shows that the stimulation of certain nerves has been successful in relieving back pain. Dr. Walker explains how the stimulation works.

Dr. Judith Walker, *anesthesiologist, neurophsiologist, Walker Pain Institute:* The electrical stimulation works by going up to the brain and at the brain some powerful substances are released — one of them is already known to be serotonin — and these substances actually go back down to the spinal cord and make the spinal cord insensitive to further pain impulses. So it actually stops the pain impulses before they reach perception. If they haven't had surgery, our success rate is phenomenal. If they've had surgery, unfortunately the success rate is lower, and it's approximately seventy percent. And the reason for that is that there's so much scar produced by multiple surgeries.

Hugh Downs: *(voice-over)* But even for people with multiple surgeries, there is hope. We visited with John Dunleavy about five weeks after his procedure. For the first time in twelve years Dunleavy is feeling better, well enough to swim and to have some fun with his kids. *(to Mr. Dunleavy)* So this is something of a miracle for you, isn't it?

John Dunleavy: It is really. It just shows in my case that surgery is not the answer — complete answer to that sciatic nerve pain.

Hugh Downs: *(voice-over)* Later that afternoon we drove to Lake Winnepesaukee, where Abby Summersgill is active again with her family. *(to Ms. Summersgill)* What if you hadn't hit the right person and the right diagnosis, and exploratory surgery had been recommended to you? Would you have gone through with that?

Abby Summersgill: I don't know. I don't like to think about it. I would have been tremendously frightened. I don't think I would have done it.

Hugh Downs: *(voice-over)* Luckily she never had to face that decision. Two weeks after her injection, she was up on water skis, and by the time we visited her, she was really enjoying herself.

Abby Summersgill: I feel totally comfortable. I guess there's a lot of force pulling me and a lot of tension on my back, but I don't feel it. I feel exactly the way I felt before. There's nothing wrong with my back anymore.

Barbara Walters: What we've shown you doesn't mean that you're to avoid back surgery at all costs. Sometimes it is clearly the thing to do. I mentioned Hugh Downs' own back problem, and he reminded me of a time during our days together on the *Today Show* when he was actually in a wheelchair because of his own aching back. Now, he had surgery for it, and it was completely successful. However, if you are not the right candidate for surgery, it's good to know that alternatives are being developed.

TRANSCRIPT 5

Dr. Jones Salk, Discoverer of Polio Vaccine

from *World News Tonight*, June 14, 1991

Peter Jennings: Finally, this evening, our Person of the Week. There are younger members of our audience who will not know him. There are middle-aged Americans and older who will recall him as an international celebrity of their youth, which reminded us how long it has been since he did so much to rid the world of a dreadful disease. We choose him this week because he has persevered and is still helping others to find the way.

Dr. Jones Salk: Immunization normally has been used in individuals who are not yet infected in order to prevent the establishment of infection.

Peter Jennings: In 1986, it occurred to Dr. Jonas Salk that AIDS researchers were becoming bogged down trying to find drugs that would cure or prevent AIDS. He began to think about another approach.

Dr. Jonas Salk: What seemed to me when I first proposed this four, five years ago that we might possibly try to immunize individuals who are already infected before the development of symptoms of disease.

Peter Jennings: Salk noticed that when a person first becomes infected with the AIDS virus that can ultimately lead to full AIDS, the body fights hard to defend itself. But over time, the body's immune system gives in. Salk began to experiment with a vaccine that would give the immune system a boost and keep the disease in check using the AIDS virus itself.

Dr. Jonas Salk: Now this was regarded as a strange idea at the time and was not looked upon with any great favor.

Dr. Allan Goldstein: He challenged conventional wisdom and he developed a new idea. He had a hunch and he stuck to it.

Peter Jennings: The announcement this week that another scientific team had taken his idea of inhibiting AIDS from spreading in the body and was testing it with some success, has been met with optimism and caution.

Dr. Anthony Fauci, *AIDS researcher:* It may not necessarily be safe, it may very well be that later on you're going to see negative effects.

Peter Jennings: Dr. Anthony Fauci, a leading AIDS researcher, is watching the vaccine's test results very closely.

Dr. Anthony Fauci: Right now in June of 1991, we haven't seen efficacy in this study. We've seen some promising, intriguing results that deserve further study.

Dr. Jonas Salk: It's always good when you've carried out, proposed a hypothesis, and obtained some evidence to seek confirmation independently on the part of others.

Peter Jennings: Jonas Salk has been in search of results most of his life. He was raised in New York. A high school teacher of his called him a perfectionist. After receiving his MD from New York University in 1939, Salk went directly into medical research at the University of Michigan where he and his team would discover the vaccine to prevent influenza. By the late 1940's infantile paralysis or polio was regarded as the modern day plague. Hundreds of thousands of children throughout the worlds were stricken, confined to crutches and breathing machines. In 1948, Jonas Salk won a grant to do research on a polio vaccine. His results in 1953 were revolutionary, a vaccine produced by a form of the polio virus itself. The same approach he has taken to AIDS. In 1955, medical researchers declared the Salk vaccine safe and effective. Millions of children were made safer. In the scientific community, Jonas Salk has not always been regarded as genius caliber. His original vaccine was very largely replaced by the Sabin vaccine only a few years later, but Salk has always tried to solve problems the unconventional, often unpopular way, and he is philosophical about his approach to the fight against AIDS.

Dr. Jonas Salk: It means that we're not putting all our eggs in one basket and as we know how nature works, it's through diversity. Soon nature will give us the answer.

Peter Jennings: Jonas Salk is 76 now, not nearly ready to retire and in the fight against AIDS he does believe he is onto something.

Dr. Jonas Salk: The end of the road is some time away, we're on a journey you might say and we're attempting to find the best way. What this experiment tells us, is finding so far, is that we must go on. Nature has not said no to this path.

TRANSCRIPT 6

Ninetysomething

from *Prime Time Live*, September 3, 1992

Diane Sawyer: I want to show you a picture tonight. You're going to see it on the cover of *Life* Magazine next week. It's the same beautiful woman, beautiful but changed by forty-eight years. The article in the magazine is all about a subject we decided to explore last year. The race to find some scientific breakthrough that will stop aging in its tracks. As of now our genes and cells seem programmed to sign off in about a hundred and ten years, a hundred and twenty, maximum. Though genetic scientists have already doubled the life of the fruit fly, for humans it's still pretty far away, if at all. So, we did a kind of survey of what's just around the corner. What makes Robert Browning a prophet for saying, "Grow old with me. The best is yet to be."

Diane Sawyer: Paul Spangler is running his fourth mile of the day — uphill, downhill. He's in training for the New York Marathon. After mile eight, he's a little out of breath. But then Paul Spangler is ninety-two-years-old.

During the week, Sadie Colar plays two or three sets a night in New Orleans clubs and hotels. That's eight to midnight. She makes her way home and then gets up at six to walk to the market to do the shopping. Welcome to the gay nineties, Sadie style.

And in this studio, a prolific potter works eight hours a day on pieces that sell for up to $30,000. After a long struggle, business is now booming for Beatrice Wood, who this year turned - ninety-eight.

[Happy Birthday to you.]

Beatrice Wood: When people ask me how I happened to live such a long time — because I'm really very old, old, old — my reply is, "Chocolate and young men," and that solves it.

Diane Sawyer: Beatrice, Sadie, Paul — they may seem like fortune's children, blessed in their old age. But here's a headline for you. Scientists are now convinced that in the next two decades, the nineties are going to be like that for a whole lot of us. They're going to feel like the back end of middle age, because every month in laboratories around the country scientists are making startling discoveries breaking open the old myths about the aging body and the mind.

Dr. Ed Schneider, *Head of Andrus Gerontology Center, U.S.C.:* Right now, I think the tremendous excitement is in the aging brain.

Diane Sawyer: Dr. Ed Schneider is the head of the Andrus Gerontology Center at U.S.C., one of a dozen places feverishly working with different chemicals to keep the brain vital.

Dr. Ed Schneider: One of the things that we have found is that the old myth that as we age, we lose brain cells, and that we can not stop this phenomenon, may be wrong.

Diane Sawyer: May be wrong?

Dr. Ed Schneider: May be wrong, that we may be able to take tired, weary, aged, damaged brain cells —

Diane Sawyer: Yes, please!

Dr. Ed Schneider: — and sprinkle some substances, one of them called "nerve growth factor", and these cells will wake up, feel terrific, and function well.

Diane Sawyer: Here's what they're doing at U.S.C. Scientists have isolated rat brain cells damaged by disease. They've then added a natural body chemical called "nerve growth factor", a chemical essential to normal brain development, and they saw incredible results. New connections were made. Fewer cells were dying. As for the rat, brain-damaged rats without nerve growth factor can't find their food in a maze, while rats with nerve growth factor have no problem. Human tests have already begun and a breakthrough could be as soon as ten years away.

So you're saying, by the early twenty-first century, people will start taking pills to change their brain chemistry.

Dr. Ed Schneider: I think so. I think, as today many people are taking their multivitamin pill in the morning, and people with specific diseases take other pills, we'll be taking pills that will allow us to age more successfully, and there may be a pill that will allow us to have the same brain power at age ninety — it'd be wonderful — that we had at age twenty

Dr. Morton Shulman: When I first took the pill, to my great amazement, within hours I was feeling better and the next day I was back to my normal state of health.

Diane Sawyer: In Toronto, Dr. Morton Shulman says, the magic pill is here now. It's another chemical for the brain called Deprenyl. Shulman took Deprenyl for the brain disease Parkinson's, and — after a dramatic reversal of his symptoms — bought the Canadian rights to the drug because he's convinced it can do wonders for healthy humans, too.

And Shulman sent us to the lab where he's funding his own research. Here, scientists put an old rat, the equivalent of an eighty-five-year-old human, in a water maze. Even though the rat has been shown where the platform is, he can't find it on his own. He's more interested in trying to get out. But when an old rat is given Deprenyl, it can find the platform at the same rate as this young rat.

But back at Shulman's office, his staff has seen all the research they need to see. They're all taking Deprenyl now, even though it's only been approved in the U.S. and Canada for Parkinson's patients.

1st Staffer: Sure, I'm a Deprenyl user. That's right.

2nd Staffer: If it can help me live a few years longer, to enjoy life, then why not do it?

Diane Sawyer: As for evidence of increased, well, friskiness —

3rd Staffer: Actually, I have seen a change in my sex drive.

4th Staffer: The woman that I'm going out with has complained on occasion that I'm too active.

Diane Sawyer: Now, wait a minute! Let's be very clear about this. There is absolutely no proof that Deprenyl does anything for healthy humans. And what works in animals doesn't always translate to people. Not only that, doctors will tell you if you're healthy, to take a pill on a gamble is crazy., At the same time, there is going to be some chemical compound, some breakthrough pill, and it's right on the horizon, a pill that will help a lot of people keep their brains as young and agile as ninety-eight year old Beatrice Wood.

[**Beatrice Wood:** Oh, that is so seductive.]

Beatrice doesn't have to wait for a pill. She does it with her spirit.

Did you ever worry about how long you would live? Did it ever matter to you?

Beatrice Wood: No, Very often, when I'm very exhausted, I think, "Oh, now this is it. One more breath and I'll be gone." And then my revolting mind says, "Yes, but you wanted to do that pot and you wanted to make that shape." And before I know it, my mind has brought me back to life.

Diane Sawyer: It's as if Beatrice wills herself into eternal youth. Just look at her pottery. She produces fifty to sixty pieces a year.

Sadie Colar isn't waiting for science to come up with a pill, either. She was twenty when she started playing piano on Mississippi riverboats and the work still keeps her brain supple, especially her memory.

Sadie Colar: I learn all the new songs, play with all the bands in New Orleans, every band you can call.

Diane Sawyer: You just pick them up by ear?

Sadie Colar: If I hear something, I can play it. Just, as long as I know what key it's written in, I play it.

Diane Sawyer: Sadie is living proof that, contrary to what everybody thinks, memory doesn't have to collapse in old age. In fact, only one percent of people under age seventy-five get Alzheimer's. Most of the news about memory is good. It turns out a lot of memory loss has less to do with age than with depression, other illnesses, side effects of pills you're taking, lack of confidence — and that's something you can reverse with a little practice.

136

Look what happened at the Memory Assessment Clinic in Bethesda, Maryland, with a sixty-six-year-old man who could only remember one out of fourteen names on a test. He practiced taking a face and associating it with imagery. And after just one week, he was scoring twelve out of fourteen, exactly the same as his thirty-five-year-old daughter.

Paul Spangler says he has trouble remembering the new songs he learns with his barber shop quartet, but by his own admission, he's not really practicing that. He spends his time — you remember — staying physically fit. He is a retired doctor — he was a surgeon at Pearl Harbor — who decided to increase his exercise dramatically, later than you'd think.

How old were you?

Dr. Paul Spangler: sixty-seven.

Diane Sawyer: You didn't really start to take care of yourself until you were sixty-seven?

Dr. Paul Spangler: I didn't care. I wasn't — I wasn't — well, I was having a good time, but I was overweight and not physically fit, I was physically active, but not physically fit, never had been.

Dr. Maria Fiatarone, *Scientist:* Our experience has been that there's really no age at which you can not make people stronger.

Diane Sawyer: Scientist Maria Fiatarone is conducting the first strength — training tests on people in their eighties and nineties at the Hebrew Rehabilitation Center outside Boston. What she learned even about people who've had strokes and heart attacks, defies the common logic.

Therapist: Are you flirting, Philip?

Patient: Why not?

Diane Sawyer: Dr. Fiatarone had discovered that a one hour workout three times a week can increase a ninety year old muscle ten percent in a matter of months.

Dr. Maria Fiatarone: Which is quite dramatic, and that's about what you'd see in a twenty-year-old, for example, who was doing weight training.

Diane Sawyer: And there's another key area of research into what and how much you eat. Roy Walford doesn't believe it's propaganda. He's the dean of diet research and these are his mighty mice, equal to ninety-year-old humans. But they've had their calories restricted to almost half their normal diet and the result: Their life span increased by almost fifty percent and they had a lot less disease. We asked Dr. Walford to compare a restricted mouse to a fully-fed one.

Dr. Roy Walford: You can tell a great difference if I hold them up together. This is the restricted one.

Diane Sawyer: Look how gray this guy is!

Dr. Roy Walford: This one is very gray. He's got material from the cage stuck to his rear end. He doesn't groom himself very well. His hind legs are not functional. And yet these are the same chronologic age. If I just stroke them, you can see which on is the more vigorous.

Diane Sawyer: He's convinced it's no accident that the humans with the longest life span are the ones who naturally eat a low-calorie, high-nutrient diet, namely, the people of the island of Okinawa in Japan. The Okinawans have the highest percentage of people over a hundred of any culture in the world. Walford also says that calorie restriction not only works in mice, but in other species like fish, cattle, monkeys. An he believes it will work in people, too.

Dr. Roy Walford: If this applies exactly to humans, it means by this technique, humans could live to be a hundred and fifty to a hundred and seventy.

Diane Sawyer: You are convinced that humans can live routinely to a hundred and fifty - to a hundred and seventy.

Dr. Roy Walford: Not on an individual basis, but the survival curve of the human population, which generally terminates — goes along like this and then gradually terminates and comes to zero at one hundred and ten, which is the genetic maximum life span limit to our species normally, this curve can be extended, not ending here but ending out here at, say a hundred and fifty. Yes, I'm convinced of that.

Diane Sawyer: Whether he's right remains to be seen, but one thing is already certain. In the future, for more and more people, the nineties will be as vital and creative as middle age. The nineties could be the best years of your life.

What age would you like to be in the hereafter?

Dr. Paul Spangler: Well, I've had so much fun at this age, I don't think I don't mind being this age.

Beatrice Wood: I know that I'm almost a hundred to you, but I'm really only thirty-two to myself, so I refuse to grow up!

Diane Sawyer: What's the best time of your life?

MS.. Colar: Well, I'll tell you, I've been having the best time of my life ever since I've been born!

TRANSCRIPT 7

Ability Grouping Can Stifle a Child's Educational Development

from *American Agenda*, September 5, 1990

Peter Jennings: Once again education is on the *American Agenda*. Our focus, as it was last night, is on a practice so widespread that most parents and educators don't give it a second thought. We're referring to ability grouping, putting kids who seem to have the same learning abilities together. The trouble is that ability grouping doesn't work. In fact, it has the potential to affect a child's total educational development and develop a scar which may last into adulthood. Our Agenda report is by ABC's Bill Blakemore.

Bill Blakemore: Is it possible that whether you become a successful manager or a prison inmate, a research scientist or an unskilled laborer, a professional artist or one of the chronically unemployed, is it possible that might be determined partly by an educational sorting process which begins when you're six years old? Yes, quite possible. New studies have found that what is called ability grouping or tracking is common in most American schools. Two-thirds of America's elementary and middle schools and more than ninety percent of high schools use some form of tracking or ability grouping. In Boston's Haley Elementary School, for example, children are sorted into fast and slow classes from the first grade on up. Parents here, like most in America, accept this.

Woman: If you mix the groups totally, you're going to lose not only the bright child or the slow child, you're going to lose both.

Woman: So they're going to teach to the middle and so, so you are going to lose both ends.

Bill Blakemore: The theory is that ability grouping keeps fast kids from being held back by slow kids and keeps slow kids from becoming disheartened when they try to keep up.

Robert Berry, *Principal, Haley Elementary School*: We're putting them there because we feel it would be of a better benefit to them, that in the end, they will achieve more.

Bill Blakemore: The theory sounds logical, but researchers have found it doesn't really work that way despite what many teachers believe.

Anne Wheelock, *Educational Researcher*: Within their year with the kids, they might see results, but they really don't see the deficits that accumulate over time. The don't see the test scores drop from year to year. The bottom tracks get bigger and bigger and bigger throughout the secondary years.

Bill Blakemore: And the high tracks get smaller. Children at every level of ability will tend to achieve less in a track system.

Teenage Boy: They usually have them categorized like A, B, C, D, and you felt lesser in the groups and you seem to only make friends in that one group.

Teenage Girl: And I felt like... felt dumb and I thought they didn't expect a lot, so why do a lot?

Bill Blakemore: Ability grouping is self-perpetuating for several reasons. The most obvious is that is demolishes self-esteem.

Jeannie Oakes, *UCLA*: If you're told that you're just average, or worse, that you're slow, you're likely not to expect yourself to be anything but that.

Bill Blakemore: Many school districts say students are free to move up from low ability groups. But the new studies show, that almost never happens. It's "catch 22." If students haven't had the right preparation for advanced classes, they can't move out of their lower ability classes, which aren't giving them the right preparation for advanced classes.

Teenage Boy: The connection was how the teachers felt about us. The teachers knew that we were in

138

those lesser groups, and so they weren't really trying like they would with an A student who didn't understand something.

Bill Blakemore: And that's not the only way students get shortchanged. In school after school, surveys have found kids grouped in lower levels usually get less qualified teachers, less access to learning resources, and less stimulating curriculum.

Dr. Philip Grignon, *Superintendent, South Bay Union School*: And so you give them the worst materials, the worst teachers, and you stick them in the worst classroom, you try to ignore them and then you call them bad and you wonder why they drop out.

Bill Blakemore: But can school systems challenge every student without the damaging effects of ability grouping, without predetermining their chances in life? The Northfield, Massachusetts school system replaced its ability grouping five years ago with a system which includes kids of different abilities teaching each other.

Dona Cadwell, *Guidance Counselor*: The college attendance has improved and kids' aspirations are way up. We don't have people who are rejected because of what level they are.

Bill Blakemore: One school district in California is developing another alternative to grouping. Individual education plans for every student. These schools are proving what some teachers have always known, that children who elsewhere would be put in low ability groups perform at high levels given the right kinds of teaching. Teaching which produces as many ability groups as there are children and does not decide for them what they'll be able to learn. Bill Blakemore, ABC News, San Diego.

TRANSCRIPT 8

Cooperative Learning:
A Good Alternative to Ability Grouping

from *American Agenda*, September 6, 1990

Peter Jennings: Because this is back to school week for so many, we again put education on the *American Agenda*. Last night and the night before we focused on ability grouping — the practice of putting children who seem to have similar abilities together. Some on the fast track, others on the slow. But, as we've already learned, educators have been finding that ability grouping does more harm than good. Moreover, it practically guarantees failure later in life for kids on the slow track. On the Agenda, we try to put forward alternatives, and as Bill Blakemore reports tonight, there are some.

Bill Blakemore: In a growing number of schools, teachers are asking the kids to put their heads together, and it's getting remarkable results. It's called cooperative learning, and teachers say it's producing higher test scores — even for slow students — trouble free school integration, fewer discipline problems. And it can do all this without sorting children into fixed ability groups. Instead of spending all their time in traditional rows, passively listening while the teacher does most of the talking, students in cooperative learning schools also spend time in small groups. The teacher sets the task, then the kids figure things out together.

Robert Slavin, *John Hopkins Univ.*: The teacher's still the best teacher in the class. A kid will never do the job of a teacher, but what kids are terrific at is re-explaining and individualizing and making material stick with each other.

Girl: I like it 'cause if there's something you don't understand you can always ask like one of your friends to help explain it better to you. Maybe you'll understand it more if they explain it to you than whether if your teacher would.

Bill Blakemore: The teachers can improve their communication, too, moving from group to group.

Teacher: I can zero in one-to-one or one-to-four.

Bill Blakemore: And do so while all kids stay engaged.

Teacher: If I'm here teaching and somebody gets stuck, they have someone there, they don't have to wait.

Bill Blakemore: They got a roomful of teachers.

Teacher: I think the greatest benefit is the kids can move at their own pace.

Bill Blakemore: No one is held back. No one gets stuck. No one is forced to move on 'till he really gets it. No one in Tai's group could explain reducing factions, so he went over to Steve. Back in his own group, Tai then taught Jenny. Some of what goes on in this new cooperative education was called cheating when those of us who are a little older were back in school — kids looking at each other's papers, talking to their neighbors.

Boy: We share our answers and...

Boy: We discuss the problems.

Bill Blakemore: Finding out from each other how to get the answers. And that's the point. Having kids explain things to each other uses what teachers have always known: you really get to know something when you teach it.

Teenage Girl: 'Cause if you teach them something then you know that your know it, and it stays in your mind longer than if you were just to stare at a board.

Bill Blakemore: What's remarkable is that this all works partly by mixing together students of different levels of accomplishment as opposed to separating them

Roger Genest, *Co-op Learning Director:* I saw students who I would have considered my top academic students suddenly being put on the spot and really having to explain what they thought.

Bill Blakemore: It encourages less advanced kids to learn because each student gets points, not for the most right answers, but for individual improvement. Those points then contribute to the group's score in classroom competition.

Lynn Marquardt, *Co-op Learning Director:* And so gradually they become a better student and they feel much better about themselves because they're more accepted by the other children in the classroom as contributing members and valuable.

Bill Blakemore: The students also know they'll be tested individually. The teachers in the most successful programs keep running records of each student in each subject, so they know when a student needs to spend time concentrating alone. And when the teachers rearrange the learning groups every few weeks, they match kids whose strengths compliment each other.

Robert Slavin: I think kids in cooperative learning get more of a feeling of, "I know how to get information. I know how to learn. I know how to solve these problems. I can do it with my buddies. I can do it myself," rather than "I'm totally dependent on what the teacher does."

Bill Blakemore: Once kids have learned they can learn for themselves, it's much less likely they'll be pigeonholed by any ability grouping system which decides for them what they can know or what they may become. Bill Blakemore, ABC News, East Allen County, Indiana.

TRANSCRIPT 9

Media Studies Would Help Kids Watch TV More Critically

from *American Agenda,* March 3, 1992

Peter Jennings: On the *American Agenda* tonight, - how to watch television. We've put the subject on the Agenda because by the time they leave school, American children will have spent more time in front of the tube that in front of the blackboard. So they ought to know how it works. Our Agenda reporter is Beth Nissen.

Beth Nissen: By one definition, these American children are illiterate. Not because they are watching hours of television instead of reading books, but because they are watching hours of television passively and not actively reading its messages.

Professor Renee Hobbs, *Babson College:* It surrounds us and yet we pay no attention to it. And we uncritically take in those messages without thinking, "What's not on the screen?" "What's not being shown?"

Professor Robert Kubey, *Rutgers University:* We permit our culture to know almost everything it knows about, mostly from television and the news media, but we give almost no education to children about where these pictures come from.

Beth Nissen: Students in other countries — Australia, Great Britain, and here in Ontario, Canada — are being taught in school to watch carefully and think critically about what flashes past on television.

Barry Duncan, *Media Studies Teacher:* What about those images? Do they disturb you?

Beth Nissen: Barry Duncan designed the media studies curriculum in Ontario, where media classes are required for all high school students.

Jen, *Student:* You never see women that are just slightly overweight or are trying to lose weight.

Barry Duncan: What we're really getting is the male point of view. If we don't pay attention to some of those things, then, in fact we're not detecting bias, we're not detecting point of view.

Teacher: It's just a shot, but it's a judgment. Here's a guy who's larger than life. What's a great way to show that? He's larger than the screen.

Beth Nissen: Students learn how use of the camera influences their perceptions; how a wide shot establishes a scene; how a low-angle shot helps give a subject greater stature and authority; how a series of quick shots grabs viewer attention. Students learn the power of editing.

Boy: They're going to show the parts that they want to, that support what they want us to see and they're going to cut out the parts that contradict that.

Beth Nissen: The students themselves say these classes have transformed the way they look at, television news, sitcoms, commercials.

Mark, *Student:* I always try and figure out where the person who made this... where they're coming from, what their angle is. What do they want me to do as I watch this? What do they want me to think?

Beth Nissen: Yet, in the United States, which produces and watches more television than anywhere else in the world, media study is taught only in scattered classrooms. .

Teacher: What kinds of words are used?

Beth Nissen: In this experimental class at the Oyster River Elementary School in Durham, New Hampshire, fourth and fifth graders watch the news, then analyze the way a story is represented.

Boy: They didn't show Tsongas and Clinton as much as they showed Kerrey and Harkin.

Beth Nissen: Students learn just how limited air time really is by editing and producing their own news stories; by struggling with what to include and what to leave out.

Girl: Two minutes and twenty-three seconds and twenty-five frames.

Jack Callahan, *Teacher:* They are faced with making decisions, And now we say there's somebody else on the other side of your television at home who's making those same decisions. How can they use or abuse that power?

Beth Nissen: Many school districts, overburdened and over budget, say they cannot afford to train teachers or buy equipment for media literacy classes. But those who promote media education say any teacher — in fact , any parent — can begin by spending more time with children while they watch television and asking them questions about what they see.

Professor Robert Kubey: You can say, "Why do you think they did that just now?" or, "Who's that?" "Are they telling us the truth all the time in commercials?" We can ask our kids those kinds of things.

Beth Nissen: It is a skill children will need to use all of their lives, as students, as workers, as citizens. How to see what everyone else sees but how to think about it for themselves. This is Beth Nissen in Toronto for the *American Agenda.*

TRANSCRIPT 10

A Dinosaur Named Sue

from *Prime Time Live*, January 7, 1993

Announcer: And now from Washington, Sam Donaldson.

San Donaldson: It may be the custody battle of the century. Scientists, Indians, the U.S. government — they're all trying to lay claim to a bunch of old bones. Well, not just any old bones. These bones happen to be sixty-five million years old. In the midst of it all is the man who dug them up in South Dakota. For him it was the find of a lifetime. Now he could wind up in jail.

Sylvia Chase has the amazing tale of the fight over a dinosaur named Sue.

Pete Larson: I started collecting fossils when I was four years old. I cannot envision doing something else.

Sylvia Chase, *ABC News:* Meet Pete Larson, a grownup with a job any kid would envy — he's a dinosaur hunter. And here's the movie version of the dinosaur of Pete Larson's dreams — Tyrannosaurus rex, a fearsome meat-eater more than three-stories high who ruled the earth millions of years ago, still a creature of mystery. Only nine skeletons have been found so far.

Somewhere out here, though, there was a tenth T. rex, and finding it would be the crowning achievement for any dinosaur hunter. Larson's native South Dakota is dinosaur country, laced with fossils of North America's prehistoric creatures.

Pete Larson: Every time you turn over a rock you might find something that no one else has ever seen before.

Sylvia Chase: Then one hot August day in 1990 …

[**Pete Larson:** I believe that the tail is going this way and the head is going that way. But we're just going to have to dig it up and see.]

Larson had found T. rex number 12, and his crew made a home video of the excavation.

Pete Larson: This is one of his front teeth.

Sylvia Chase: They named the dinosaur Sue after paleontologist Sue Hendrickson, who led them to the bones.

Pete Larson: The skull is right here, lying more or less right side up. This is where the eye would be and the cheekbone, the lower jaw.

Sylvia Chase: The dinosaur named Sue had been buried in that hill over there for sixty-five million years. It's in a remote corner of a ranch in central South Dakota. She may be the greatest find in the history of dinosaur hunting — the biggest, the most complete and the most valuable.

And so it was headline news, and Pete Larson was on top of the world.

Pete Larson: Because of the discovery of this one skeleton, we're going to open a new chapter in the book of dinosaurs.

Dr. Bob Bakker: Sue is the finest Tyrannosaurus rex ever found.

Sylvia Chase: Leading dinosaur experts like Dr. Bob Bakker confirmed the importance of Larson's find.

Dr. Bob Bakker: Sue is giving us an insight into how this last dynasty of dinosaurs. There were many, many, many, many ages of dinosaurs, and the very last was ruled by Tyrannosaurus rex. We want to know about that ruler. Why did it rule? How did it die?

Sylvia Chase: Larson says Sue died at about a hundred years of age and he thinks he knows how, because he found the tooth of another T. rex in Sue's facial bones.

Pete Larson: She probably died as the result of a bite from another Tyrannosaurus rex. The whole left side of her face seems to be just torn away.

Sylvia Chase: But the most important thing about Sue is that ninety percent of her was recovered. She's more complete than any of the existing T. rex skeletons.

Pete Larson: We have the first complete tail. For the first time we know how many vertebrae Tyrannosaurus had in their tail. We have the first complete shoulder blade. We have the first complete arm.

Sylvia Chase: And the first virtually complete skull — all five feet of it. And for the first time, stomach contents — a duck-billed dinosaur believed to have been Sue's last supper.

Pete Larson: The amount of knowledge that we can obtain from this specimen is just staggering.

Sylvia Chase: Larson's company is expert at the incredible, delicate task of separating fossils from the rock encasing them. With customers like Harvard, the Sorbonne, and the Smithsonian. Larson's Black Hills Institute is located in tiny Hill City, South Dakota, population 600, but it's reputation is worldwide. Do you make a lot of money in it?

Pete Larson: No, I don't make a lot of money.

Sylvia Chase: Yet, even after carting off perhaps the most valuable T. rex in the world, Larson announced it would go into a non-profit museum in Hill City, the kind of place he had in mind since finding his first fossil. So Sue was not for sale.

Pete Larson: Sue is a premiere piece. Why would we sell the best piece, the piece that's going to bring thousands and thousands of people to visit that museum?

Sylvia Chase: That was great news in Hill City, a tourist town where, even with a nightly western gunfight and an 1890 train, the average income stays at $3,000 below the poverty line. The hope was the dinosaur museum would stop vacationers who roll through town on their way to Mount Rushmore.

Mayor Drue Vitter: We were dreaming high and big at that time.

Sylvia Chase: Hill City Mayor Drue Vitter.

Mayor Drue Vitter: Can you imagine little-bitty Old Hill City going to national recognition in a year or two's time, where people said, "Have you been to Hill City?", when they didn't even know it existed before?

Sylvia Chase: But the dream of Hill City and Pete Larson turned into a nightmare on the morning of May 14th.

Pete Larson: One of our workers came over and said, "The whole place is crawling with FBI." And I said — my mouth dropped open — I just didn't know what to say.

Sylvia Chase: Armed federal agents had shown up to seize the bones of the dinosaur, saying they'd been stolen. What was going on? Well, it turns out Sue is worth is as much as $20 million, and when word got around, you can call it envy or just plain greed, but suddenly people started claiming Sue was theirs.

Who owns Sue?

Maurice Williams: If the laws of the land mean anything, I own it.

Sylvia Chase: Who owns Sue?

Kevin Schieffer: The public owns Sue.

Sylvia Chase: Who owns Sue?

Pete Larson: The Black Hills Museum, of Natural History Foundation.

Sylvia Chase: Larson says he owns Sue because, when he found her on the 35,000-acre ranch of Maurice Williams, he paid Williams $5,000 — top value for a fossil still in the ground.

The Williamses say the sale doesn't matter and they own Sue because they're Indians, and Sue was removed from a section of the Williams ranch which is protected by a trust agreement with the U.S. government.

Maurice Williams: Well, I told him more than once that he should clear it with the federal government, and he knew that he should. He chose not to.

Sylvia Chase: And now that they know what Sue's worth, the Williamses want their cut of that twenty million.

Isn't it the money?

Maurice Williams: It's always the money.

Mrs. Williams: It's always the money. That's what it's about.

Sylvia Chase: Twenty million dollars would also mean a lot to the Cheyenne River Sioux tribe. The average income here: $7,000 a year; the unemployment rate: eighty percent. For Indians, looting of artifacts from reservations — including fossils — is a long, seething issue.

Tribal chairman Gregg Bourland contends that though Sue was taken from the privately held

Williams ranch, the ranch lies within the reservation boundaries.

Mr. Bourland: All of this land has restriction to it, and you cannot just say, "Go ahead, dig up a big old dinosaur. Here's five grand, I'm out of here."

Sylvia Chase: At one point the tribe proposed a dinosaur partnership to Pete Larson, but negotiations broke down when Larson insisted he had to be sole owner.

Mr. Bourland: I gave my hand of friendship to Mr. Larson. The Institute, they slapped it away. That's a hand that I will think twice before I extend it again.

Pete Larson: I think that if there was a wrong done by us, it was by perhaps not keeping in better touch with the tribe.

Sylvia Chase: At the tribe's urging, the U.S. government stepped into the fray and, if things weren't complicated enough, U.S. Attorney Kevin Schieffer declared, quote, "The fossil is the property of the United States." Period.

Mr. Schieffer: It is a public treasure. It is a national treasure.

Sylvia Chase: And so calling the bones criminal evidence, Schieffer ordered the federal agents to crate them up and take them away. Sue's bones are now locked in a steel box at the South Dakota School of Mines.

Pete Larson: I felt like we were being raped. We were being — these people were here stealing from us. These are the people who are supposed to protect us. These are the police who are supposed to keep things like this from happening, and here they were perpetrating this crime.

Radio DJ: It's 7:40 taking the Black Hills by storm, just like the FBI did. Here's "Hill City Sue" at KKLS.

Singer: *Hill City Sue / Hill City Sue / The government stepped in / Oh, gee what else is new?*

Sylvia Chase: Throughout the area, protesters, angered by the loss of big bucks that Sue represented, demanded the dinosaur be returned to Larson.

Pete Larson: What right does the government have? That fossil would have never been found without us. It would never have been excavated without us. It would have never been prepared without us. It'd be sitting in the ground and washing away, year by year, until she rots away.

Sylvia Chase: The feds didn't see it that way. So though Larson had paid $5,000 for Sue and spent $100,000 to restore her, the government was now saying Larson stole Sue and other fossils from federal land.

U.S. Attorney Schieffer says policy prevents him from talking about Larson's case, but he indicates it involves more than just the excavation of Sue.

Mr. Schieffer: If you're going on federal land and stealing federal property, and if you're going on Indian land and stealing Indian property, that's reprehensible, and that's got to be prosecuted.

Sylvia Chase: But who owns what's out here? Some areas are a patchwork of Indian land, private land, federal land.

Pete Larson: They're looking for the chance that I was in a pasture, with which I had permission to go on, and perhaps went across an invisible boundary and picked up something from somebody's property that I didn't have permission for.

Sylvia Chase: In the case of Sue, Larson invited scientists from around the country to come look, which is not something a thief would likely do.

Mayor Vitter: If you and I stole a Rembrandt, we sure wouldn't tell the public we had it. "Come on in and see what I stole. Examine it. See if it's authentic." We would hide it.

Sylvia Chase: But sources told *Prime Time* Larson will be indicted on major felony charges. He's already gone into debt in his fight with the government.

Could you be driven out of business because of this?

Pete Larson: Oh absolutely. The government, in cases like this, wins their war through attrition. It would take and use up every last one of our resources. And they — you know, it's bottomless pit there with the government — they have — they're using our money to fight us, too. They're using tax money to do this.

Sylvia Chase: Are you willing to tell the U.S. taxpayers what you've spent so far?

Mr. Schieffer: I am staying within my budget.

Sylvia Chase: Meanwhile, the dinosaur named Sue is staying in the lockup, still in that steel box where Larson says she could turn to dust.

Pete Larson: Sue is a living thing, and she needs to breathe. Her bones need to breathe in order to not decompose. They've taken the most fantastic fossil in the world and they put her into boxes where she's being destroyed.

TRANSCRIPT 11

Joey's Best Friend

from *Prime Time Live*, January 30, 1992

Diane Sawyer: We've always heard that dolphins are our sole mates under the sea and you may have read some of the studies of what they can do. They can listen to a human voice and mimic it with their own sounds. They can memorize different configurations of seven or eight symbols, the same way that we remember telephone numbers. But it's one things to learn that dolphins have intelligent brains and another to learn they have sympathetic hearts, if you will - human kindness.

John Quinones traveled to the Florida Keys where he found a kind of magic rapport between handicapped children and dolphins: a phenomenon described by sociologist Betsy Smith.

John Quinones, ABC News: *(voice-over)* For more than fifty million years, long before man ever walked the earth, they have graced this dark underworld with majestic beauty. For decades we've known them as the most intelligent creatures of the sea, next on the intellect scale to humans and primates. It now turns out they may be smarter and perhaps more sensitive than we thought. Scientists at Dolphins Plus, a research center in the Florida Keys, have discovered that dolphins display a surprising array of human like feelings, like joy and compassion, particularly for children who are physically or mentally impaired, children like five-year-old Joe Hoagland.

Deena Hoagland: Never ever did I expect a baby to come this way or to be hurt this way or to have to suffer in that way.

John Quinones: *(voice-over)* Joey was born with truncus arteriosis, a rare congenital heart disease that robs the body of oxygen. By the time he was three years old, Joey had undergone three open heart surgeries. After the last operation, he suffered a stroke. The left side of Joey's body was paralyzed. He was blind in his left eye. Deena and Peter Hoagland could only watch as their infant son lay in a coma for eight days.

Deena Hoagland: He was absolutely limp. He was less than a newborn. He didn't have swallow response. He couldn't talk. He couldn't cry. There were times when — two times that I can remember — that the doctors came in and said, "We don't think that you should leave. We think that maybe he's not going to make it."

John Quinones: *(voice-over)* Joey finally came out of the coma. Doctors said he might even recover, but after such a long ordeal he refused to cooperate with doctors or physical therapist.

Deena Hoagland: The kid was petrified. Can you imagine yourself being in a coma and waking up to something like that, with bleeping machines?

John Quinones: He didn't trust humans.

Deena Hoagland: I was concerned he wouldn't trust us.

John Quinones: *(voice-over)* The Hoaglands had just about given up hope for Joey's recovery, but then they heard about a natural dolphin habitat in Key Largo, Florida

(on camera) Dolphin Plus is a research center, not an entertainment resort, so there are no shows here where the dolphins perform tricks for the crowd. In fact, here the dolphins are free to swim out to sea if they choose. And while they're taught to respond to hand signals, they don't have to earn their food by performing. Whatever interaction there is with humans here is strictly at the dolphin's discretion.

(voice-over) Although the emphasis is on research, the center does allow a select group of properly trained visitors to swim with the dolphins. It was at one of these sessions thirteen years ago that Betsy Smith, a sociologist at Dolphins Plus, noticed that her mentally retarded brother showed improved mental agility after swimming with dolphins. After a bit more experimenting, she discovered the dolphins worked especially well with children.

Betsy Smith: As soon as I put a handicapped child in the water, this dolphin relaxes, becomes very different — very, very quiet and calm, and will stay with and work with that child as long as it takes.

John Quinones: *(voice-over)* By the time Joey arrived at the center last year, the twelve dolphins who live here had helped dozens of children.

Betsy Smith: That's part of what we call their "altruistic behavior." They care for their sick and they care for their ailing and their elderly. And we think it's an extension of what's natural behavior for the dolphin. You communicate, you care, you live in "synchronicity", we call it. They're being dolphins.

John Quinones: *(voice-over)* When he first laid eyes on the dolphins, Joey was shy and apprehensive. But then a friendly dolphin named Fonzie swam up to greet him and though Joey was afraid to get into the water, Fonzie would not take no for an answer. A few days later, the swim therapy began. Joey weighs 40 pounds. Fonzie 700 pounds. Yet when Joey couldn't move his leg, the powerful dolphin nuzzled up and gently moved it for him. When Joey couldn't lift his arms above his head, Fonzie would toss out a ball and make Joey toss it back. Pretty soon Joey, who couldn't unclench his paralyzed hand, was practicing in his sleep so that some day he could grab a fish to feed Fonzie.

Deena Hoagland: I heard him talking, so I went into his room and he was fast asleep going, "Try again. Open, shut them. Open, shut them. Just keep trying and you'll get it."

John Quinones: What did Fonzie do that medicine couldn't provide?

Chris Blakenship: Emotional therapy.

John Quinones: *(voice-over)* Chris Blakenship is a marine biologist and Joey's therapist.

John Quinones: But this is a dolphin.

Chris Blakenship: I think that there is an aspect to dolphins that no one yet really understands.

John Quinones: Exactly what does Fonzie understand about Joey's illness? Well, no one knows for sure, but some experts believe dolphins can sense when a person is ill. Already some studies show that dolphins display a powerful stress-reducing effect on children. Kids will swim with them, become deeply relaxed. and therefore much more receptive to teaching.

Dr. John Schull: These animals are not just unfeeling swimming hunks of meat.

John Quinones: *(voice-over)* Doctors John Schull and David Smith are scientists and experts on animal behavior. They've just completed a two-year study on the mind of the dolphin. They taught a dolphin named Natua to respond to different pitches of sound underwater as seen on this videotape. Watch the three panels at the bottom of the screen. When Natua heard a high-pitched tone, he learned to push the panel on the right. When he heard a low tone, he pushed the one on the left. But then the researchers made things a little more difficult on Natua by playing tones that were neither high nor low.

Dr. David Smith: He starts to make a lot of errors and gets a little upset with us during that phase of the experiment. But at that point we give him a third panel that he can press, which he can press to escape into an easier trial.

John Quinones: And that, say the researchers, was their most important finding. Natua knew when to say, "I don't know."

Dr. John Schull: We know he doesn't know the right answer yet. And the fact that he uses his bail out response on those trials suggests to us that he know he doesn't know the right answer yet.

John Quinones: Why is it so important for us to know that?

Dr. John Schull: If you realize that these are thinking, feeling, perhaps self-aware animals, you realize that they have a much deeper kind of relationship with humans.

John Quinones: *(voice-over)* A relationship with humans, the researchers say, is greatly enhanced by the dolphins' remarkable sonar capacity.

Dr. John Schull: They can perceive very fine variations in hardness and texture and size, using their hearing alone.

John Quinones: So they can read a human being in the water, for instance.

Dr. John Schull: They can read a human being in the water and given what we know about sonar, there's reason to suspect that they may be able to use their sonar to perceive things that are going on within the skin, the same way that physicians now use ultrasound in order to get a look at fetuses in the womb.

John Quinones: *(voice-over)* That sensitive sonar, say scientists, may allow dolphins to detect certain handicaps in humans. That comes as no surprise to Deena Hoagland, who's convinced that Fonzie knows exactly what he's doing for Joey.

Deena Hoagland: Something very wonderful happened. They developed a friendship and Fonzie accepted Joe just the way he was and showed Joe what he could do. Fonzie never concentrated on what Joe couldn't do.

John Quinones: *(voice-over)* After just eight months of therapy, Joey now lugs a heavy pail of fish to feed his buddy.

Joey Hoagland: He took it right out of my hand!

John Quinones: *(voice-over)* Once he couldn't move a finger on his left hand. Now he wiggles it away, for that's the signal that makes Fonzie talk, dolphin-style.

Joey: Kowabunga!

John Quinones: *(voice-over)* He still takes daily doses of medicine to keep his heart beating properly.

Deena Hoagland: It's to relax the muscle in the heart.

John Quinones: *(voice-over)* But aside from the medicine, Joey seems as normal as any other five-year-old.

Deena Hoagland: There were doctors that said he would never run. There were doctors that said that he would always walk with a limp. There was a speech therapist that told me he would never sing.

John Quinones: What has Fonzie done for you?

Joey Hoagland: He helped me feel better.

John Quinones: *(voice-over)* A few weeks ago, the boy who once couldn't walk jumped into Fonzie's world for the ride of his life.

Deena Hoagland: I'm always in awe that this is my little boy who's with a 700-pound dolphin. Their relationship is kind of like two best friends that are secret pals, that share something so very special that they'll both know about for the rest of their lives and will never ever forget.

TRANSCRIPT 12

Environmentally Friendly Design

from *American Agenda*, August 26, 1993

Diane Sawyer: We've put the workplace of the future on the *American Agenda* tonight. In Chicago this summer, 5,000 architects from all over the country signed what they called a declaration of interdependence — a pledge to design buildings that are kinder to the environment. As our Agenda reporter Barry Serafin tells us, some buildings like that already exist.

Barry Serafin: City skylines are impressive. But across the county, a growing number of architects say buildings like these waste energy, trap polluted air indoors, and add to environmental problems instead of solving them.

Randolph Croxton, *Architect:* We have just become accustomed to accepting really poor buildings

Barry Serafin: But that is changing.

William McDonough, *Architect:* We are looking at a cultural shift as significant as that of the industrial revolution

Barry Serafin: It is called "green architecture", and environmental groups are helping to lead the way.

This is the new headquarters of the National Audubon Society in New York. Instead of building a new structure, architects recycled an old one, saving hundreds of tons of steel and concrete and 9,000 tons of masonry.

Architect Randolph Croxton was told that every design decision had to make economic, as well as environmental, sense. For example, sunlight pours through skylights and windows. Electronic sensors brighten and dim lights, depending on the amount of daylight.

Sensors also detect when offices are empty and when they are occupied. The lighting system cost an extra $100,000, but is twice as efficient as in most offices.

Randolph Croxton: We have a pay back in energy alone — $100,000 a year — that will very quickly recoup any margin of additional investment.

Barry Serafin: Green architecture means making more use of recycled materials and trying to make buildings healthier. So floor tiles here are made of old, crushed light bulbs. The carpeting is natural wool with no dye or glue to give off chemical fumes. And the air is cleaned and recirculated six times an hour.

Jan Beyea, National Audubon Society: I think what's happened here is we've taken several hundred small little details, put them all together, and made a great building. This is a great place to work. It feels wonderful.

Barry Serafin: Finding environmental building materials can be difficult but suppliers are beginning to sprout up. For example, a company called "Environmental Construction Outfitters" has assembled 2,600 new products and technologies — everything from carpets made of recycled plastic jugs to insulation made of cotton from recycled blue jeans. Architect Paul Bierman Lytle founded the company.

Paul Bierman Lytle, *Architect*: We really want to get this out of boutique stores for the environment, but into the lumberyard.

Barry Serafin: The new ideas are already beginning to show up in surprising places. There is no city more Middle American that Lawrence, Kansas.

And there is no retailer more mainstream than Wal-Mart. But this new Wal-Mart store is different from the almost 1,100 others across the country, starting with a parking lot made of recycled asphalt.

Inside the Lawrence store, there is an environmental education center. But the store itself offers the real education, with skylights and energy efficient lighting, benches made of recycled plastic and bins to collect more for recycling.

Outside, waste water is collected in a holding pond, treated, and then used to water the native shrubs and grasses used for landscaping. Wal-Mart often outgrows its stores within ten years, so this one has been designed so it can someday be recycled and turned into an apartment house, office building or factory.

Wal-Mart officials will not say how much more the new experimental store costs, but it is thought to be ten to twenty percent more than their other stores. They do say ideas tested here may be used in future stores.

For advocates like Susan Maxman, president of the American Institute of Architects, green architecture is already beyond the experimental stage.

Susan Maxman, *Architect*: It's not a style. It's not a fad. It is there. It is our survival as a species, obviously, and as a society.

Barry Serafin: Green architecture is no longer a novelty. With a wide array of new technologies, it is becoming more available, more cost effective, and further, harder for architects and the companies that hire them to ignore.

Barry Serafin, ABC News, Lawrence, Kansas.
